# DO
## *EVERYTHING*
## BETTER
### WITH
## MUSIC

# REVERBERATION

## KEITH BLANCHARD
## FOREWORD BY PETER GABRIEL

**ABRAMS IMAGE, NEW YORK**

# CONTENTS

# HERMANN'S HEAD

A NOTE FROM THE EDITOR

It was a typical Saturday night in 1978. Folks were out, my big sister, Lisa, and her friend Susan Higgins were on babysitting detail, and I was a little pest. "Paradise by the Dashboard Light" blared from Lisa's room, and I was able to sneak in to witness the commotion.

The epic song had reached that familiar verse—the Phil Rizzuto baseball part—and the girls proceeded to provide me with a way-too-young-to-know, base-by-base life lesson I've not soon forgotten. *"...He's rounding first and really turning it on now, he's not letting up at all, he's gonna try for second..."* There I was with a big, stupid smile, soaking it all in.

It was a great first track for my own personal playlist. It never felt so good, it never felt so right.

Lisa passed away a few years later, but left me a most precious gift—Styx and Zeppelin, Foreigner and Andy Gibb, The Police and Peter Gabriel, keeping her spirit alive and launching my love of music.

Now, lifetimes later, I'm learning to appreciate my music in a whole new way, using learnings from neuroscience to leverage my favorite songs to improve my life in multiple ways: helping me focus, stimulating my creativity, calming me down, helping me sleep. My new journey is all about uncovering how music's magic happens—and it's a hell of a ride.

Joining me on this awesome journey is Peter Gabriel, himself—a giant WOW! for me—and his talented daughter, Anna. We're working tirelessly to get to the bottom of all this: to discover why our brains love music so much and how, in fact, it can help us do everything we do, well...better.

The book in your hands is the follow-up to our first brain endeavor, *It's All In Your Head*, and plants our flag in the field of music—sort of a debut track kicking off what we hope will ultimately be a lengthy playlist of memorable hits.

So stay tuned as we develop this core idea into a world of multi-channel entertainment. There's a lot more detail to follow.

Come join us on this brain train, and learn how to do everything you do better with music.

**- MICHAEL HERMANN**

# FOREWORD
## BY PETER GABRIEL

I've always thought of my music collection as a box of pills—a toolkit for transformations. Different music serves different goals: Music can be used to take us into dance, battle, sport, ritual, sex, serenity. It can change how we function, in groups or as individuals, how our minds and bodies work, how we feel and see the world around us. Even how we see ourselves.

Once, I jumped off a tall stage during the last number of a set. A wild dancer had crossed into my line of descent, and we collided, and something snapped in my leg. I was lifted back onstage, and finished the rest of the number on my knees. The band walked off, but I remained there, unable to move—they thought I was just hamming it up. I had to be lifted off the stage, and my leg began to hurt like hell. But what was really interesting to me was that, even though I knew something was wrong, I had felt no pain at all in the last three minutes. The adrenaline provided by the music, the crowd, and performing had completely anesthetized my broken leg.

The more research that emerges, the more powerful music appears to be. In this book, we will explore the special relationship between music and the brain, the engine that processes these vibrations, sounds, and harmonies and turns them into all manner of stimuli and action. Of all the senses, sound seems to go through less mental filtering before it manifests in the body (with the exception of language, which seems to follow a much more circuitous route). Low frequencies can vibrate our bodies directly, and the rest of the frequencies seem to coax out specific feelings without a lot of mental effort. We try to make sense of all the inputs, create order out of chaos, work out who we are and where we need to go, and find some meaning in the cacophony that bombards us.

Pattern recognition appears to be one of the principal functions of the brain, and the ability to synchronize your brain to a musical rhythm is present across cultures. When we listen to music, we are detecting and anticipating its form, trying to find a match with anything else in our memories and social filters. Repetition of sound—in rhythm, harmony, melody, or words—can hold a special power. How many times do you do something before it gets boring...and can you transcend that boredom and use the repetition to take you somewhere else?

We used to have a house in Senegal, where music and dance are still as much a part of everyday life as food. I saw a mother teaching her baby, who could barely stand, how to dance. She was clicking her tongue while her baby moved in time with a big smile, flapping his arms and bouncing up and down. Although I am not known for my skills as a dancer, I was always being invited into the center of the dance circle. In Senegal, being an awkward white man is no excuse for not dancing, singing, or playing—it's just what you do. There is so much freedom in surrendering to the power of music and allowing it to take your body and mind to places you don't normally go.

When writing songs, I am very conscious of what feelings I am aiming to create, and occasionally I will have a very specific goal. A song like "I Grieve" was intended to give people a tool to help them come to terms with loss. I had been looking for something that could help me, and I hadn't found what I was looking for. In most spiritual traditions, a search for such meaning is associated with silence: a voluntary choice to look for more with less. Does a sound mean more when served in silence, like a drop of water in a pool? Context is another important factor when trying to unravel these mysteries of music and mind.

We are entering an age of big changes: biomonitoring, genetic manipulation, artificial intelligence, and—potentially the most disruptive of all—the BCI, or brain-computer interface. Whether we connect directly to the circuits of the brain or access them noninvasively, extraordinary things are already happening as this new frontier opens up. The ability to read, write, and translate brain activity is about to turn the world upside down.

It will also help clarify how the sensory inputs activated by music can be engaged to change our behavior. Many years ago, we did a show called *The Lamb Lies Down on Broadway*, and my plan for the beginning was to take brain and body readings from each member of the band and turn them into music. It was 1974, and the technology wasn't yet able to deliver what I had imagined. Today, it's all there—and more. If we choose, we can all become the creators of our own self-generated sound and light show, which, using some smart AI, we could learn to design ourselves to serve our needs at any time. Bringing AI into the musical mix will allow us to turn our own brain activity into self-generated music: less deejay, more "me"-jay.

We all have different ways of interacting with music, and for many of us, listening is just something we do without a lot of thought, like breathing. But if we can start to understand this mercurial stuff called music a bit better, it might give us a powerful toolkit to deploy whenever and however it is needed—music as medicine, as educator, as therapist. This book is not going to provide all the answers, but I hope it will allow us to ask better questions.

# WHATEVER DON'T SKIP THE INTRO- DUCTION

**LET'S GO BACK TO THE ZYGOTE MOMENT.** When you were a sperm and an egg, together—nothing else. You were you, already, but there was still much to be done.

For months and months, before your grand entrance into that frenzied alternative universe of the hospital delivery room, your cells were furiously dividing. You were a piece of glitter, then a grain of rice, then a penny. Your spider veins multiplied; you sprouted nubs that morphed into paddle hands and flipper feet. And at some point, that gigantic embryo head of yours started to become...aware.

But what was it that sparked your consciousness? It surely wasn't something you *saw*: Your eyes do open in the womb, but not until about your 28th week, and frankly, there isn't much to see. It wasn't a smell or taste either, there in your undifferentiated, fluid-filled sensory deprivation tank. No, what lit the fire of your consciousness—what woke you up, so to speak—was

# YOU DO,

unquestionably a sound. Because at 18 weeks or so, before your other senses truly came online, you started to be able to hear.

And *that* is where things get interesting.

It would have gone something like this: There is nothing...and then, slowly, there is something. A steady, pounding rhythm. *Bump-bum, bump-bum, bump-bum.* It fills the whole space; it is the noise the world makes. You can't know what it is yet, of course, but it seems important—and it is.

This is your mother's heartbeat: the very first song in the unique and wonderful soundtrack that is your life.

Welcome to the show.

It's hard to overestimate the importance of a mother's heartbeat and voice to a newborn. Doctors and nurses lay a baby back on Mommy's chest as soon as they can: face to face, skin to skin, putting their heartbeats back in proximity. Those heartbeats don't sync up precisely: Infant hearts beat about twice as fast as Mom's. But for the newborn child, in this horribly loud, too-bright new world, encountering that familiar mother-sound again, outside the womb, serves as a beacon of continuity in a world gone mad. And presto: Bonding begins.

These sounds no doubt remain a baby's favorite music for a long, long time. But the role they play is not merely about comfort: It turns out that those early audio inputs are a fundamental part of the newborn's continuing physical development. Specifically, the auditory cortex, where sounds are processed and elevated to the level of perception, doesn't fully develop until about week 24 of a pregnancy. And since we have long had the ability to save premature babies born before this, we could theoretically test, outside the womb, just how critical these inputs are to that development. And because we could, we did.

Researchers in Boston, led by a Harvard neuroscientist, divided extremely premature babies into two groups. One group received standard neonatal care. In the other group, each baby heard, for three hours a day in the incubator, recordings of their mother singing "Twinkle, Twinkle, Little Star" and reading *Goodnight, Moon*, as well as a reproduction of her heartbeat. Thirty days later, the researchers discovered that the babies who'd received these extra inputs had thicker auditory cortices, a phenomenon correlated with better hearing and language development.

**CORPUS CALLOSUM**
*is a nerve bundle connecting your brain's right and left hemispheres*

That's right: The sounds of their mothers literally altered the structure of these infants' brains and made them stronger. And that singular amazing idea—that sound can be used as a tool to make meaningful changes to your brain—sets the stage for everything you're about to read.

From the very first moments that you are physically able to hear a beat, music is beginning to shape you. As we'll see again and again throughout this book, music isn't merely entertaining: It continually reshapes your ever-changing, "neuroplastic" brain...for better and for worse. Year after year, new studies reveal more specific benefits of strategically played music. It can help you sleep well and wake up refreshed, focus on challenges, reduce **stress**, unlock real creativity at that brainstorming meeting, work out your body and your **emotions**, connect with others and reconnect with your past—and a whole lot more.

**AMYGDALAE**
*(one in each hemisphere) help with decision-making and emotional response*

**MIDBRAIN**
*processes rudimentary vision and hearing*

**PONS**
*is involved in breathing, sleeping, swallowing, and similar involuntary activities*

This book is about using science to leverage music's powers to improve your life in every possible way. But before we can start to comprehend how and why music exerts this incredible, hypnotic influence over our brains and bodies, we have to review the basics. So let's start with the miracle of your ears.

**MEDULLA**
*regulates breathing, heart rate, coughing, sneezing, and other unconscious activity*

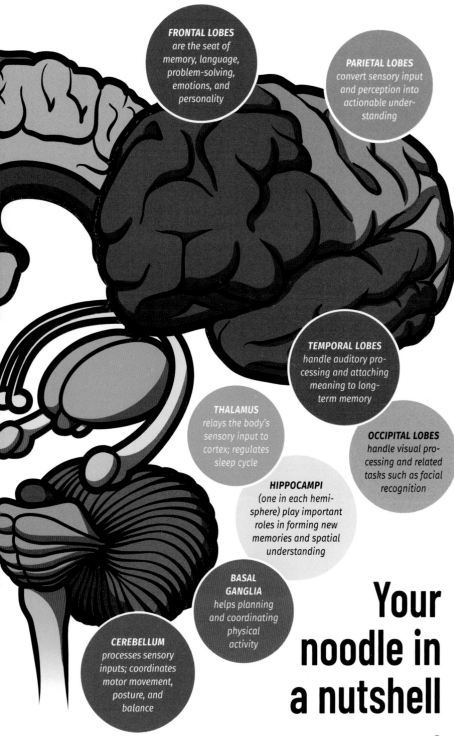

**FRONTAL LOBES** are the seat of memory, language, problem-solving, emotions, and personality

**PARIETAL LOBES** convert sensory input and perception into actionable understanding

**TEMPORAL LOBES** handle auditory processing and attaching meaning to long-term memory

**THALAMUS** relays the body's sensory input to cortex; regulates sleep cycle

**OCCIPITAL LOBES** handle visual processing and related tasks such as facial recognition

**HIPPOCAMPI** (one in each hemisphere) play important roles in forming new memories and spatial understanding

**BASAL GANGLIA** helps planning and coordinating physical activity

**CEREBELLUM** processes sensory inputs; coordinates motor movement, posture, and balance

# Your noodle in a nutshell

# Meet Your Brain Waves

*Your brain's general level of activity is characterized, for convenience, into five frequency ranges, from very slow—let's call it your mind's "first gear"—all the way into overdrive. Here's what happens at each stage.*

### First Gear: Delta Waves (.5–4 Hz)
This is the lowest level of brain activity, occurring when you're unconscious and your brain is essentially on autopilot, mainly just keeping your body's background systems running. Delta waves happen during Stage 3, dreamless sleep (as detailed in Chapter 1: Relax), and when you're deep in a meditative state. Inside your head, this is as quiet as it gets.

### Second Gear: Theta Waves (4–8 Hz)
Theta waves describe the slightly higher level of brain activity that occurs when you're in that gray area between being asleep and awake—either sleeping lightly (e.g., REM sleep), or technically awake but couch-bound and metaphorically comatose (e.g., watching bowling).

### Third Gear: Alpha Waves (8–12 Hz)
If your brain is generating alpha waves, you're fully awake, calm, unworried. This has been called the resting state of your brain: It's where your head is when you're showering or standing in line, when you're driving straight but your mind is wandering.

### Fourth Gear: Beta Waves (12–35 Hz)
This is your brain's active state, generated during the productive part of your day, when you're wide-awake, alert, and focused. When you're producing beta waves, you're paying attention to sensory inputs, having conversations, and interacting with the world around you.

### Overdrive: Gamma Waves (35+ Hz)
Gamma waves are generated when your mind is at its busiest. You're actively processing information, hyperaware, solving problems. You may think you're multitasking—a myth we dispel in Chapter 2: Focus. Are you having a knock-down, drag-out fight with your boss? Racing the clock while taking the SAT? Welcome to gamma time.

Imagine you're in the middle of a very loud movie theater, having a low-pitched conversation with your significant other while the movie's dialogue, sound effects, and stirring soundtrack simultaneously stream from speakers all around you. You hear a particular kind of hollow clatter, which you instantly recognize as the sound of a dropped popcorn bucket, about three rows behind you. And when someone from the far side of the theater murmurs, "Butterfingers!" and a titter runs through the crowd, because they also understood what fell and why, you hear that as well—without missing a bit of the movie.

All those sounds expressed themselves to you as a single, complex, invisible pressure wave, experienced by both ears, which your brain had to process, split into individual impressions (the movie's layered soundtrack, your local conversation, the clattering, the audience noise), and resolve into a credible story of what's happening around you. Your ears and audio processors not only tease out individual sounds but attribute direction and distance, measure threat level and intuit cause, compare to memory and predict outcomes, and a lot more. And it all happens in real time, at a mostly unconscious level, in the pitch dark.

Here's how your ears work their magic—the four-part, backstage journey that sound makes on its way into your head.

## THE OUTER EAR

Your outer ears may seem like pointless trifles, an evolutionary afterthought slapped on to keep your sunglasses from falling off. But they're really mini satellite dishes that amplify even the faintest sound waves so predators can't sneak up on you and prey can't get away. (Cup a hand around the back of an ear right now and you'll double its size...and power.) Having two ears lets you triangulate incoming sounds so you can understand where they're coming from, but for simplicity's sake, let's focus on one ear. The sound wave captured by the outer ear charges on down the fleshy tunnel of your ear canal, where it ultimately bumps up against a stretched-skin boundary at the end that scientists call the tympanic membrane, but which you know affectionately as your eardrum. The resulting vibration is felt on the other side of this thin, rubbery wall—the pressure-sealed chamber known as the middle ear.

## THE MIDDLE EAR

What's inside your middle ear? Only a brilliantly synchronized jazz trio made up of ossicles, the tiniest bones in the human body, nicknamed for their distinctive shapes: the Hammer, the Anvil, and the Stirrup. This trio—hell, let's call them The Ossicles—plays continuously in these cramped quarters, vibrating sympathetically with whatever's going on outside the eardrum wall. When birds chirp and raindrops patter and Adele belts one out, it's The Ossicles, your in-house cover band, that faithfully re-creates the sounds here inside your head. These "sounds" are still pressure waves that the brain can't interpret, mind you, but they're now occurring in a protected, pressure-controlled, lint-and-earwax-free environment. They won't become electrical signals the brain can use until they reach the inner ear.

# THE INNER EAR

The Ossicles' faithful cover version of the sounds of the outside world is played for an audience of one: the cochlea, a fluid-filled, curving snail shell whose inside is lined with tens of thousands of tiny hairs. The distinctive pattern of each sound wave, as it travels into the curve of this shell, triggers unique combinations of these filaments, converting the pressure wave into tens of thousands of individual electrical impulses that head out, via the **auditory nerve**, and up into the brain for processing.

# THE AUDITORY CORTEX

This teeming multitude of individual hair-trigger signals heads first to your right and left auditory cortices for untangling. Pinch the four fingers of each hand together and tap them against the side of your head, an inch above your ears— this is where your auditory cortices are. They sort sound by **frequency**, time (so you hear the gunshot before the ricochet), and other basic factors, and they contain subunits that handle more complex processing, such as **Wernicke's area**, which helps us understand language.

When you listen to music, it's The Ossicles' version that's digitized and sent out to be simultaneously reviewed by a number of different stakeholders in your brain: the memory-focused hippocampus (Do I know this song? Should I sing the lyrics?); the **motor cortex** (Should I tap my feet, or get up and dance?); the moody amygdalae

(Does this song make me sad?); and so on. All these parallel reviews are then combined, with the results tied up neatly into a compelling story, essentially in real time, by your busy frontal cortex. You recognize this song...you like this song...let's crank this song up...let's dance!

This process—the recording, rerecording, **encoding**, mixing, and reviewing—goes on continuously, every waking moment of your life.

But wait: There's more. Auditory inputs aren't merely impressions to be passively translated and "understood" by our all-knowing brains. Rather, sounds and other sensory inputs actively adjust your brain—and through the brain, your body and behaviors—whether you're conscious of these adjustments or not. The brain, alone in its silence and darkness, wants to know what to do based on what's happening in the world. It's hungry for feedback; it responds actively to cues from the real world. And that is a powerful opportunity you can leverage.

The specific mechanism by which you can use cues from music to influence your brain and behavior is complex and fascinating.

Your busy brain never stops whirring and working, even when you're asleep, or in a coma, or nodding off while a coworker documents his fascinating dream. But the patterns of electrical impulses sweeping across your **synapses** do change over the course of your day, in ways that correspond to your behavior. (See Meet Your Brain Waves, page 6.) With advanced tools like EEG and fMRI visualizations, we can measure these waves of electrical activity and represent them as visual snapshots that demonstrate how active each region of your mind is at any given moment.

As your activities change over the course of the day, your brain can seamlessly switch frequencies, most often between resting alpha and busy beta states. We don't always get the state we want or need, as you already know: Your mind can race along annoyingly when you're trying to fall asleep, or it can slip inexorably toward slumber when your boss is droning on in the conference room. The ability to harness the power to control your brain's activity level would surely be a game changer.

Wouldn't it be amazing if you could actively induce the wave pattern you need at any given moment? To relax deeply or to focus intently, to supercharge your workouts and inflame your passions, to improve your connections with friends and family, to induce productivity or creativity on demand?

As you may have guessed by now, it turns out you can. ■

# RELAX

# SLOW DOWN, YOU GOT TO MAKE THE MORNING

**SIMON AND GARFUNKEL**
"The 59th Street Bridge Song
(Feelin' Groovy)"

# YOU MOVE
# TOO FAST

# LAST...

**WHEN I CATCH AN OLD MOVIE OR TV SHOW,** I'm always struck by just how slowly life used to move. In the black-and-white rendering of life from just a few generations ago, nobody has more than one job—and the one they have includes a long, boozy lunch and ends promptly at five. People take the time to dress up for airplane flights. They spend all morning with the newspaper; they stroll down to the barbershop, if the notion strikes them, to get their face professionally shaved. Dates or families sometimes just "go for a drive" to nowhere. If our culture's fiction is to be believed, life was slow, and simple, and sweet.

Well, needless to say, it moves a hell of a lot faster these days. Instead of going for a walk, we work out while watching our favorite shows. We take long weekends instead of two-week vacations. We mute ourselves for video conferences so other participants won't hear our keyboards clattering as we listen with one ear while cheating with other work. And any spare moments that happen—those times when people used to look around, or say hi to a stranger—we fill instead by interacting with our smartphone, in case anything irresistibly fascinating happened in the 43 seconds since the last time we checked it.

If we're going to explore how we can leverage music's extraordinary power over our brains, relaxation is undoubtedly the place to start, because we already have an instinctive understanding of how it intersects with music. You may not yet use music to improve your workouts, strengthen your relationships, tap into creativity, or stave off **dementia**, but we all know how to use music to relax.

> "I been waiting for the waves to come and take, take me right to you."
>
> POST MALONE "Otherside"

Imagine you're setting up a playlist for a happy hour with friends. (One of the great advantages of life in the 21st century is the effortless fingertip access to all the great music of the world—everyone's a deejay now.) My playlist, like yours, is a unique reflection of my personal tastes, my age, my history, my friends, my life. I'd probably throw in some Steely Dan and Elvis Costello from my college years, some Blondie and Boz Scaggs from high school, some guilty-pleasure favorites I hope my friends won't mock me for, and so on—an intensely personal mix of songs that I know, instinctively, will get my friends of a certain age quickly into the chill zone.

Which specific songs should I choose, to bring the vibe? The lyrics can sometimes help: Songs that are great for relaxing and drinking and chilling out with friends are often literally *about* relaxing and drinking and chilling out with friends.

But mostly it's just about choosing songs at the right pace, with the right associations, played at the right volume. A well-chosen set establishes the mood quickly, shifting my arriving guests' minds into a relaxed state so they can leave their daily stress behind and get to where I want to take them. But how, and why, does this work?

Your brain activity can be measured in terms of its electrical wave patterns. When you feel relaxed, you're producing low-frequency alpha waves. This is a good, nontaxing, healthful mindset to spend a lot of time in: Alpha-wave production has been correlated with reduced stress and **anxiety**, with aiding mental coordination and mind-body integration, with improving your ability to learn, and with promoting both calmness and—oddly—alertness. Alpha waves are also associated with a boost in creativity, which we'll cover in more detail in Chapter 7: Create.

But it's not easy convincing our busy brains to downshift into the alpha state. In fact, studies have shown that our brain's activity actually *increases* when it doesn't have much to do, though the science isn't clear as to why. Some propose that our brains use downtime to catch up on important but not pressing challenges, like addressing unsolved problems or untangling complex social issues. Others say different elements of the brain seize the unhurried moment to practice working together—like your frontal cortex fine-tuning its relationship with your **fight-or-flight** reflex, to help you panic only when appropriate—or that we use downtime to organize memory, both by making sense of our recent past and by proactively planning for the future. (What does planning for the future have to do with memory? Neuroscientist Moshe Bar of Bar-Ilan University speculates that daydreaming about the future lets you lay down "prior experiences" you can draw on as you go, like when you can't stop your brain from rehearsing a stressful conversation you intend to have later.)

Relaxation is an ongoing challenge for our species, perhaps because we haven't been at the top of the food chain for long. For delicious fleshy animals like us, evolution rewards vigilance: Those who stayed constantly aware and awake, endlessly evaluating fight or flight options, were the ones who survived, by and large. We evolved for a more dangerous world than we find ourselves living in today: We're naturally jumpy, both quick to panic and slow to set stress aside when the danger passes. We desperately need the relief of relaxation—we just have to find a way to get there.

## HOW MUSIC HELPS

Ask any baby: Music can be an incredibly powerful tool for lulling humans into a state of relaxation. The sound drowns out distractions, the familiarity renews pleasant associations, and the melodies and harmonies demand partial attention from your know-it-all **prefrontal cortex**,

replacing or reducing the pernicious influence of stressors. But those are all just fringe benefits. The real show, the dominant influence of music on your brain, is **rhythm**.

Music with a strong beat can induce your brain to synchronize with its specific rhythm and mirror it, producing brain waves of the same frequency. When you listen to relaxing music—music with a slow **tempo** of 60–80 **beats per minute (bpm)**, like "Perfect," by Ed Sheeran (63 bpm), or "Hotel California," by the Eagles (75 bpm)—the pattern infiltrates your skull through the ears and the auditory nerves, and your pattern-hungry brain locks onto this and synchronizes with it to start producing alpha waves (8–12 Hz), which make you feel more relaxed. It's as simple as that.

"**Harmony** and vibration that is perceived as pleasant or coherent can affect brain waves and put you in a state of relaxation," says Dr. Helen Lavretsky, a professor-in-residence in the psychiatry department at UCLA and an expert on integrative mental health. "This state can further promote physiological changes by reducing stress hormones and stress response, relaxing breathing, and having profound effects on the **autonomic nervous system** as well as decreasing inflammation."

**Spontaneous synchronization**—in this case, between your brain and an exterior rhythm it experiences—may sound like pseudoscience. But we see it all the time in nature, for example in startled flocks of birds that change direction as one, or in the simultaneous flashing of lightning bugs in twilight during mating season. It even happens in mechanical systems. The great mathematician Christiaan Huygens, who in his spare time invented the pendulum clock, discovered that the pendulums of two clocks hanging from a wooden bar on his wall would soon mysteriously synchronize. It's not magic, of course: Huygens surmised, and Georgia Tech physicists confirmed 340 years later, that there's a mechanism by which a lower-energy system can sneak energy away from a higher-energy system until they're either exactly in phase or exactly out of phase. In the case of the clocks, the energy-transfer mechanism turned out to be tiny vibrations in the wooden bar the clocks were hanging from.

FUN FACT

# Adults who listen to 45 minutes of music before going to sleep report having better sleep quality, from the first night onward.

In brains, this spontaneous synchronization is called **brain wave entrainment**. When you start tapping your fingers or your foot to a song, that's your motor cortex at work—think of it as a sort of metronome that helps your brain's activity patterns sync up with this external

## How can I relax my nerves before an important meeting?

*Your brain, sensing danger, is likely operating at a high frequency—great if your plan is to run away or physically attack your coworkers, but not the state you want to be in to project quiet confidence in the conference room meeting you're leading. To bring your brainwave frequency down quickly, you need to give yourself five full minutes before the meeting. Find a quiet spot and disappear into your headphones with a calming and familiar song, one on the slower side and with a simple melody. (One music therapist recommends Adele's "Someone Like You" as a good example; just make sure it's one that works for you, with no negative associations.) While it plays, breathe slowly. Get lost in the song and forget everything else. At the end of the song, take one deep, cleansing breath, breathe out slowly, and head in. You got this.*

rhythmic input. Remember, at the concert, when you stood next to the speaker wall and felt the drumbeat in your chest, and you feared the sound was influencing your heartbeat? You were half right: Music realigns your mind, too.

Why would your clever, otherwise well-protected brain allow itself to be so easily manipulated by outside influences? Turns out it's a feature, not a bug. "Our brains depend on gathering information from our environment," neuroscientist Dr. Daniel Levitin, author of *This Is Your Brain on Music*, pointed out to me. "Nervous systems become increasingly sophisticated, but they still need to be permeable so that you can find food, and a mate, and avoid danger, and such." Your brain has evolved to continually absorb a hot mess of sensory inputs, decide what's important—is there an emotional component? a correlate in memory?—and respond instantaneously. By manipulating influential inputs like music, you can gain a measure of control over your brain from the outside, so to speak.

# WHATEVER YOU DO, DON'T THINK ABOUT ELEPHANTS

Slower brain waves are associated not just with relaxing but with hypnotic and meditative states. When you meditate, your brain generates more alpha waves (particularly in the posterior of your brain, closer to the **brain stem** and the seat of your older, "reptilian" brain functions). Because we're so distractible, it's hard to get into a meditative state at will, never mind staying there—meditation is a discipline that's typically practiced and learned over a long period of time. But it stands to reason that music in the slow, 60 bpm range known to induce alpha waves should be able to help practitioners shut out the world...and so it does. "When I'm meditating, I'm taking out all the drums," says EDM artist Steve Aoki. "I want to be floating. I want to be levitating. And in order to be in that kind of mental state, you need to have music that gets you to that place."

**Binaural beats** are an interesting channel for inducing a meditative state. Freely available as sound files on YouTube and elsewhere, binaural beats send music of a slightly different frequency to each ear, and your brain perceives the difference in frequency as a third sound. (The effect works better with headphones, naturally.) When that third sound's frequency is in the target range for meditation... *voilà*. One controlled study showed that listening to binaural beats significantly reduced anxiety in emergency-room patients, an anxious group if ever there was one.

Relaxing music can provide nontrivial health benefits, like lowering blood pressure and keeping one's heart rate down. But it doesn't have to be music as it's traditionally defined. Tibetan singing bowls, for example, can reduce anxiety, tension, and **depression**, which are all known precursors to various kinds of disease.

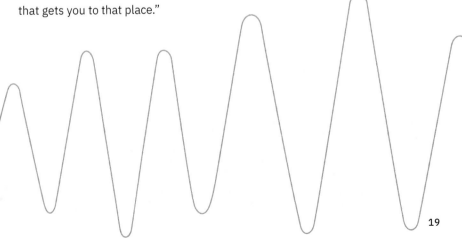

FOUR QUESTIONS FOR

# DAVID BYRNE

There is only one David Byrne. The exuberant founder of pioneering new-wave band Talking Heads has crossed boundaries and platforms, leaving a distinctive legacy in theater, in film, in visual art, and, of course, in music.

**Are there times where you'd just rather not listen to music?**

Lots of times. I suspect that as someone who writes music, performs it, I hear music in a more analytical way, and I either have to completely ignore it or really give it my attention. And so I rarely listen to music, unless I can give it my attention. So I do it when I'm traveling, I do it in the evening, if I'm doing something that I don't have to give much attention to: doing dishes or peeling potatoes or whatever it might be. For that kind of thing I can put on music, I can really get into it deeply then, but just having it on in the background, I can't do it. It either sucks my attention from whatever I'm doing or I have to completely ignore it, which is sometimes hard to do.

**How do you think technology has affected the way we listen to and experience music?**

I've noticed that for people younger than myself, they're much more catholic in their musical tastes than, say, people of my generation. I like to think that I am, but I've noticed that a lot of younger people listen to a broader scope of music—music from a 30-, 40-year span. They don't make judgments based on, "Oh, that's uncool. It was my parents' music," or that it's uncool because that person wore really kind of terrible clothes or that person is considered totally uncool at the moment. They will make judgments based on what they hear. And if they like it, they like it. And it doesn't matter where or when it comes from. And I find that really encouraging. I think with streaming, the fact that they have access to so much music and the fact that the music doesn't come encumbered with any context, it's kind of music without a context.

**What is it about the magic of synchronization?**

Most nights I'm performing. I'm doing a Broadway show that has a lot of my songs in it, and talking with the other singers, we often talk about "Where are we gonna take a breath?" You're gonna synchronize where you take breaths. And that puts you in sync, your breathing is essentially in sync. It's like yoga, where it's very intentional breathing. It's not just like, "Breathe whenever you feel like it." You have to breathe when you have that pause between the words, that's your shot, take your breath. And now you have to control how you let the air out, how fast you let it out, because you've gotta get to the end of the phrase, all that kind of stuff. I don't think about that so much when I'm writing, although occasionally I write something and I go, "Oh, that's impossible to sing."

**Is rhythm the most important element when you're writing music?**

One thing I noticed is that there are a lot of rhythms that make us move, that affect our bodies, that are the result of two different rhythms happening at the same time. It's not just one single beat, like "bump, bump, bump, bump," which would be techno music, people dance to that. But in a lot of cultures, I think we are very much moved when it's two different rhythms overlapping. One of the most common is six beats to a measure, and four beats to a measure. If you map those on top of one another, you get a kind of funky beat, and it doesn't have to be all six beats or all four beats, it can be a combination of the two. The implication to your ear, to your brain, to your body is that these two rhythms are going on simultaneously. And it becomes this very physical thing where part of your body is moving to one thing, part of your body is moving to another thing. You're not thinking about it. You're not conscious of it, but I think that really kind of nails something in your body's perception of a rhythm.

## Why do lullabies work?

*Science has confirmed what mothers have known for thousands of years: Lullabies soothe cranky babies. Simple enough to hold a baby's limited attention (one singer, no instruments, an exaggerated melody), lullabies entrain the infant brain to produce the alpha waves that slow the heartbeat and respiration and encourage sleep. The language in which they're sung isn't important—it's all gibberish to the target audience—but babies seem to remember songs they heard while in the womb. Lullabies provide long-term developmental benefits for babies and reduce parents' stress as well.*

Sounds from nature, particularly water sounds, have also been shown to induce relaxation. Your brain has evolved a "threat-activated vigilance system" to rouse you from slumber when trouble's afoot, with built-in chemical responses to danger sounds like sudden, piercing cries. That same system responds in the opposite manner when the inputs are reassuringly regular background sounds, like water flowing or waves gently crashing. As Orfeu Buxton, a professor of biobehavioral health at Pennsylvania State University, explains on *Live Science*: "These slow, whooshing noises are the sounds of non-threats, which is why they work to calm people...It's like they're saying: 'Don't worry, don't worry, don't worry.'"

# (LAST NIGHT) I DIDN'T GET TO SLEEP AT ALL

Short of a coma, the most relaxed state a human can achieve is sleep. You've spent a full third of your life under the covers, but just because it's familiar doesn't mean it's well understood. Why on Earth do we need sleep? What happens after our eyes close that's so important it's worth completely shutting down for eight hours in a row, and spending so many collective years in this vulnerable, unproductive, catatonic state?

We don't completely understand sleep yet. But we know

what happens when we don't get enough sleep, and it's not good. Sleep-deprived nurses make more mistakes on hospital shifts; night shift workers have more traffic accidents on their way home. In 2012, a sleep-deprived JetBlue pilot had a "mini-psychotic episode" in flight, leaving

# "It's not required for human survival, yet it seems to be hardwired into our brains that we need music."

DR. CHARLES LIMB *on the importance of music to humanity*

the cockpit and racing through the cabin ranting about 9/11 and Jesus; he had to be subdued by passengers. According to the American Heart Association, your risk of a heart attack goes up 24% in the week after Daylight Savings Time goes into effect in the spring (when you lose an hour of sleep), and goes down by 21% in the fall when you gain back that hour of sleep. Sleep is serious stuff.

The average person needs around eight hours of sleep per day, according to the National Sleep Foundation. But the amount needed decreases over the course of your life: Newborns sleep around 14 to 17 hours each day, while senior citizens can get away with just seven to eight hours in total, including napping at the bingo table.

The fact that you need a little less sleep

after your developmental and formative years is an important clue as to why sleep is so very important. Dr. Rafael Pelayo, a sleep specialist at the Stanford Sleep Medicine Center, told me to think of sleep not as zombie time but rather as a necessary rebuilding phase our bodies and brains demand. "We tend to think of sleep in two main categories," he said. "One part has something to do with restoring things—a metabolic restoration—and another part is involved with the development of the brain. A young child sleeps more hours because their brain is growing more."

Your brain waves, like your heartbeat, slow way, way down when you're asleep...but not to zero. While you're sleeping and unconscious, parts of your brain are still active, using the downtime free from sensory input to do things like conducting necessary background repairs (like road crews that wait until three in the morning to avoid inconveniencing traffic). Your muscles relax all the way; your digestion kicks into high gear. You consolidate the memories of the day—the chaotic reevaluation of events against known background experiences that gives us dreaming. And you use sleep to process emotions and make sense of how you feel. All these functions are important throughout your life, but they are mission-critical when you're young and your body and mind are still in development.

Teenagers are notoriously bad sleepers: Their still-growing bodies and brains need about nine to 10 hours of sleep per night, but they only get around seven, on average, which can wreak havoc on their developing brains. Sleep starts to become a problem again for many people as they hit middle age. "The real midlife crisis is sleep," Dr. Joy Allen, chair of the Music Therapy Department at Berklee College of Music, told me. "When people hit 40, 50, or 60, sleep patterns can start to become a major issue, whether it's falling asleep or staying asleep. We see a large increase in the use of Ambien."

# I'M GONNA GO TO SLEEP, AND LET ALL THIS WASH OVER ME

Falling asleep and waking up happen thanks to an endless back-and-forth battle between brain chemicals. All through the night while you're asleep, a **neurotransmitter** called **acetylcholine** builds up slowly, eventually reaching a tipping point that wakes you up in the morning. Meanwhile, your level of **adenosine**, a neurotransmitter that makes you sleepy, is low in the morning but builds up over the course of your waking day, gradually overwhelming the perky acetylcholine and making you feel increasingly tired.

Your **circadian rhythms** play a role, too: Daylight keeps your pineal gland from producing **melatonin**, another sleepy-making chemical, but when the sun goes down you start pumping out melatonin—and getting drowsy. This natural cycle is affected by a number of things: Coffee, for example, blocks your adenosine receptors, keeping you from knowing how tired you are.

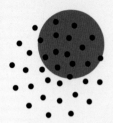

### What's white noise?

*Sound occurs at different frequencies: a low bass note, a high piccolo. And just as "white light" contains the full spectrum of light, white noise contains all audible frequencies and can help mask particular sounds of any frequency. It's used to counteract tinnitus (a medically serious, persistent ringing in the ears) and help infants sleep, and is even added to some ambulance sirens to mask background noise, enabling motorists and pedestrians to more easily pinpoint the location of the siren.*

FUN FACT

"Weightless" by Marconi Union has been found by one study to be the world's most relaxing song. It brought about a 65% reduction in anxiety.

Ideally, we want to nod off quickly, then sleep deeply and for a long time, then wake up refreshed. Over the years we've solved many challenges to this plan: You likely sleep on a mattress now instead of a flat rock, and in a dark, temperature-controlled room that's free of predators. (Interestingly, when you're sleeping in a strange place, one hemisphere of your brain stays partly awake, as if keeping watch for new dangers.) Yet reliably sleeping well remains a challenge for nearly everyone, including new parents, high school students, travelers, and six out of seven dwarfs.

Even before we leverage music, we can improve our odds of getting a good night's sleep. Keep the blue light of phone and TV screens out of the bedroom so you don't trip your brain's photosensitive circadian clock into thinking it's daytime. Invest in a firm mattress and comfortable sheets. Block out or counteract distracting noise with natural sounds that reassure your mind. Taken together, these practices are called sleep hygiene, an easy and (except for the mattress) inexpensive way to boost your well-being.

## HOW MUSIC HELPS US SLEEP

When you're trying to sleep, any soothing sound—a fan, a "white noise" generator bringing a babbling brook into the bedroom—helps eliminate obtrusive outside noises. But the right music can go much further, rhythmically guiding your brain into producing the wave type that helps you fall asleep.

To get to sleep, your best bet is to play songs whose tempo is 60–80 bpm, which roughly matches a sleeping heart rate, at not too high a volume. Instead of headphones, which wouldn't survive midnight thrashing, try sound pillows or bedroom speakers, and loop the

# THE FOUR SLEEP STAGES

**STAGE ONE** is your drowsy, half-awake, half-asleep phase. It lasts only one to five minutes, and doesn't include time lying awake thinking about that thorny work problem—this is when you're actually drifting off, and your brain's executive function (the mindset that allows decision-making) is punching out, ceding control of your consciousness.

**STAGE TWO** starts easing you into genuinely valuable sleepytime. This stage lasts from 10 minutes to an hour, and during this time your body's heartbeat slows from about 80 to 60 bpm. You're still relatively easily roused, still prone to awaken if your spouse slams the closet door or an ambulance drives by your house.

**STAGE THREE** is deeeeeep sleep, when your brain waves slow significantly, into the barely-there delta frequency. It lasts only around 20 to 40 minutes, typically, but this seems to be when a lot of the work gets done: Your muscles relax all the way, your body recuperates, your brain sorts through the previous day's memories and decides what to keep and what to lose. If it was very hard to wake up this morning, you likely were in this phase.

When we sleep long enough,
we go through these four
sleep cycles, in order, a couple
of times each night.
**Here's what happens.**

**STAGE FOUR** is the famous REM sleep, where you dream about losing your religion. Most of your body is temporarily paralyzed—except your eyes, darting about behind your closed lids, and everything necessary to support autonomic activities like breathing and digestion. And it's a good thing, too, because your brain is more or less awake, imagining sensory inputs where there are none, and combining them with memory fragments to weave together a long, rambling story you know as "dreaming." (This can happen in any phase, but it mostly happens here: If you remember your dream, odds are you woke up during this phase.)

The REM phase is critical to good sleep; this semi-active state seems to be where memory **consolidation**, emotional memory processing, and the linking together of related memories take place, while your mind is awake but disconnected from your body. If that chemical paralysis is broken, it can result in activities similar to sleepwalking, in which your zombie body tries to act on the crazy stuff your mind is coming up with. The REM phase gets longer toward the end of the night—it might be only ten minutes on your first cycle but up to an hour the second time through—which may be why you need a full eight hours to get the full benefit of REM sleep.

same song over and over, so it's familiar and soothing but requires less and less focus as the minutes wear on. What type of music you play while sleeping is of course a matter of personal taste, but it's worth noting that classical music has been shown to improve sleep quality and reduce depression in students, perhaps in part because it has no distracting lyrics.

And when the night is over, waking up the right way is important, too. One study by the Royal Melbourne Institute of Technology suggests that waking to a song provides better morning alertness than shocking your brain awake with an alarm or other annoying sound. "We think that a harsh 'beep beep beep' might work to disrupt or confuse our brain activity when waking," said the researchers, "while a more melodic sound like The Beach Boys' 'Good Vibrations' or The Cure's 'Close to Me' may help us transition to a waking state in a more effective way."

Finally, you don't need an Apple Music subscription to leverage music at bedtime. In a recent TEDx talk, longtime insomniac drummer Jim Donovan explained a simple 30-second technique he developed for putting oneself to sleep through rhythm. Sit on the edge of your bed, close your eyes, and breathe slowly, gently drumming your palms on your thighs. Keep your breathing slow throughout the exercise, but drum relatively quickly at the start (with the tick-tick-tick cadence of a stopwatch), then slow this pace down as you approach the 30-second mark. "Your brain notices that there's a pattern, it connects with it, and it begins to follow it," says Donovan. In other words, you're entraining your brain to slow down.

Sweet dreams are made of this. ◼

# TAKEAWAYS

*Relaxing and sleeping are critically important for good health. But the world is a distracting place, and the mind is slow to slumber. Along with other relaxing tools, like mindfulness and breathing exercises, music can quickly and definitively help you step off the fast track by inducing the alpha waves that slow your brain to just the right range of activity needed to relax and refresh. Relaxing more (including sleeping better) will declutter your mind, pave the way for inspiration, and confer outsize effects on your health and well-being. Something to sleep on...*

**TO SET THE TONE FOR A CHILL EVENING** Create a pleasing "sound bath," where one song plays throughout the space as a uniform background texture (as opposed to a local experience you walk in and out of) featuring chill music, in the 60–80 bpm range, just loud enough to fill the silence but not distract from conversation.

**TO TAKE SOME TIME OFF AND MEDITATE** Consider unfamiliar, drum-heavy international music, without distracting lyrics or emotional associations, to bring your brain into the alpha state and hold it there.

**TO FALL ASLEEP QUICKER AND SLEEP BETTER** Reduce light, keep phones out of the bedroom, and soundtrack your sleepytime with familiar music that's not too loud, doesn't have lyrics, and falls in the 60–80 bpm "lullaby" range.

**TO WAKE UP REFRESHED AND ALERT** Alarm clocks don't help make you alert— they just rattle you. Set your alarm to wake you with gentle, melodic music: not beeping alarms, and definitely not those super-hilarious morning radio jockeys.

# FOCUS

# I'M IN LOVE WITH MY CAN'T WAIT TO

**BILLIE EILISH**
"my future"

# FUTURE

# MEET HER

**YOU ARE AN EXPLORER ON A STRANGE PLANET,** and you have just come face-to-face with a large alien creature. Survival depends on deciding *very* quickly if this is a threat or an opportunity. Is this strange animal a predator that hunts and eats other creatures? Or is it prey, hunted and eaten, and therefore not a threat to you?

Your best instant guess is to look at the eyes.

Eyes on the sides of an animal's head provide close to a 360-degree field of vision; it's what allows tasty prey like cattle and sheep to spot the wolf coming from any angle. Eyes in the front, by contrast, help predators stay focused on whatever they're chasing so it doesn't elude them. Picture a cheetah racing across the savanna, its gaze never wavering from the wide-eyed, side-eyed gazelle out ahead, running for its life. There are exceptions, naturally, even here on Earth, but it's a solid rule of thumb: *Eyes in front: I hunt. Eyes on the side: I hide.*

It's all about focus.

## Why do some people whistle while they work? Does it help?

*Before the popularity of radio and other freely available music, groups of workers would often whistle or sing together to pass the time or express camaraderie, or as an outlet for protesting their conditions. Music was piped into America's factories during World War II to keep morale up, and the rhythm was believed to help workers stay productive. The philosophy's right there in Snow White's lyrics about sweeping up around the house: "It won't take long / When there's a song / To help you set the pace..."*

As the apex predator on our planet, humans have turned focus into an art form. We aim our forward-facing eyes into the natural world, decide what we want, and direct our energies toward developing new capabilities to obtain those things. This provides us with an ever-expanding set of tools, weapons, and other inventions: We can now outrace that cheetah, fly like birds, see in the dark, kill or heal at scale. We've spent about a million years (so far) evolving odd, bulbous forebrains uniquely adapted to rapidly process the objects of our attention. This focus on focusing, if you will, has given us agriculture, industry, culture, and supercities, opened up global telecommunication and space travel, and placed us firmly in charge of our world.

However...

Focus is becoming a grave challenge today. Our complex modern life saddles us with new cares and considerations, and supplies infinite distractions that short-circuit our goals. Maybe you want to quit your job and start your own business, or flip a house for profit, or write a

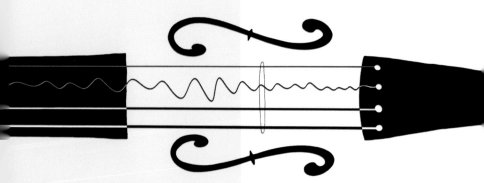

screenplay, or develop that game-changing invention. And you're going to get right on that, seriously, after maybe just one more episode of *Rick and Morty*.

We are each granted about 4,000 weeks of life—a sobering thought. Within that time span we all hope to accomplish worthwhile things, and we know that sustained focus is required to complete all the small steps that lead to big success. But our big brains wander by design, and in a world bursting with bright, shiny objects, it can be next to impossible to rein in the restlessness and stay on target.

So let's take a closer look at the fine art of staying focused, and see what answers science—and music—can provide.

## WATCH CLOSELY NOW

The human brain is a more impressive supercomputer than anything we've ever concocted, capable of processing an incredible 11 million bits of information per second. But human consciousness can deal with only around 100 bits per second—still a large number, but less than 0.001% of the totality we perceive. Our central processing unit, to adopt a loose metaphor, has to make hard choices every second as to which tiny fraction of sensory input is worth raising to the level of conscious awareness. This great winnowing defines reality for you: It helps you tune out the surrounding chatter when you're in a busy bar with a friend, permits you to safely ignore all those colorful cars on the fast-moving freeway (until one wanders into your lane), and generally get on with your life.

In other words, your first level of focus happens before you're even aware of it. Having *déjà vu,* or suddenly feeling afraid but not being sure why—the so-called "gift of fear"—can be a sign that you're reacting to inputs your brain received that simply never reached the threshold of your consciousness. ("Subliminal" literally means "under the threshold.")

Armed only with that tiny sampling of the world—the bits your brain allows you to experience—you make all your conscious decisions, including what's worth worrying about and what can be safely disregarded. And this is where we run into challenges, because modern life has grown in complexity much faster than our ability to keep up with it all.

Consider stress, which has benign origins as a useful response to emergency situations. During fight-or-flight crises, your adrenal glands, riding like cowboys on top of your kidneys, release the hormone **cortisol** to help your body cope; this improves your survival odds by temporarily slowing functions like digestion, immune system response, and reproductive system processes so you can pour more energy and resources into fighting the

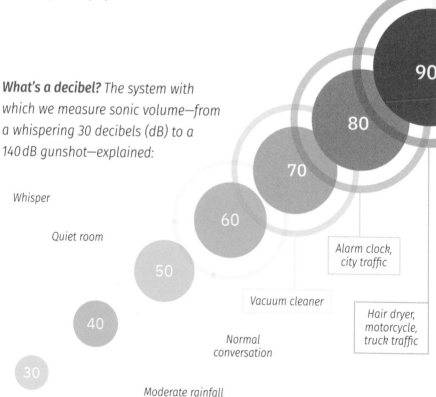

*What's a decibel?* The system with which we measure sonic volume—from a whispering 30 decibels (dB) to a 140 dB gunshot—explained:

Whisper

Quiet room

60

50

Alarm clock, city traffic

Vacuum cleaner

40

Normal conversation

Hair dryer, motorcycle, truck traffic

30

Moderate rainfall

90

80

70

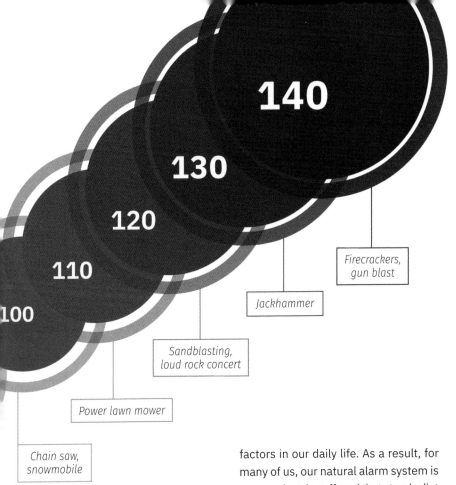

140

130

120

110

100

*Firecrackers, gun blast*

*Jackhammer*

*Sandblasting, loud rock concert*

*Power lawn mower*

*Chain saw, snowmobile*

bear or running from the bear. It's a powerful adaptive trait that helped ancient humans survive and thrive.

But in the modern world, it's not bears that cause stress—it's bear markets. And mercurial bosses. And that vaping idiot driver encroaching on your lane, from page 37. Today, stressors aren't occasional existential threats worth pausing everything to address—they're constant factors in our daily life. As a result, for many of us, our natural alarm system is constantly going off, and that steady diet of cortisol (most of the cells in your body have cortisol receptors) can lead to disastrous health consequences, including digestive issues, sleep trouble, anxiety and panic, an inability to concentrate, and heart attacks and **strokes**.

So that's one challenge to focusing: Our environment produces a heightened baseline of concern. But we're also increasingly swamped with delightful distractions, which are equally deadly for concentration. Our brains aren't simply thrown off course by distractions—they evolved to *prefer* them. As Chris Bailey, author of *Hyperfocus*, points out in a TEDx Talk, "It's not that we're distracted; it's that our brains are overstimulated. It's that we crave distraction in the first place."

Turns out humans are endowed with a perverse mechanism psychologists call the *novelty bias,* through which our brain rewards us with a hit of **dopamine** every time we take our eyes off the ball. Think about that for a minute: Your brain gets a tiny bong-hit of a chemical reward each and every time you stray from that work assignment and switch gears to check your email, or look in on the game, or try to see whether you can find Carhenge in Google Maps. (Yes, that's a thing.) It's frankly a miracle we get anything done at all.

Your brain does have a mechanism specifically designed to help us tune out distraction; in fact, it's thought that damage to this circuit may underlie the trouble many **autism** sufferers have coping with overstimulation. But it's like bringing a spoon to a gunfight. According to a recent Microsoft study, people online today spend just 40 seconds doing one thing before jumping off to something else (and drinking in their dopamine reward for the novelty). Some busy screen addicts in California, hoping to reset their dopamine receptors and break this addiction, engage in something called **dopamine fasting**—voluntarily restricting activities like sex, good food, and time with friends. It's very California, sure—but the struggle is real.

So, given that our treacherous brains overload us with false positives to worry about and reward us every time we get distracted, how can we possibly hope to focus?

Music to the rescue!

## HOW MUSIC HELPS

Handily, music addresses both of these challenges. Let's start with stress. Most people find some kinds of music to be innately pleasurable and calming. Listening to music and performing music (especially singing) have been shown to decrease both the physiological and the physical symptoms brought on by stress. Music reduces blood flow to your "fear center" **amygdalae**, which in turn lowers the production of stressy cortisol. And music increases dopamine levels, so you feel generally happier and less concerned. So far so good.

On the distraction side, music can be a powerful silencer. Studies show that ambient music at around 70 **decibels** (about as loud as a vacuum cleaner) is sufficient to muffle the intriguing human chatter around you in your cubicle farm of an office. This effectively isolates your brain from outside audio inputs, helping you to focus, although if it gets louder than about 80 decibels, that music starts to become interesting itself, and therefore intrusive. Workers at some Starbucks

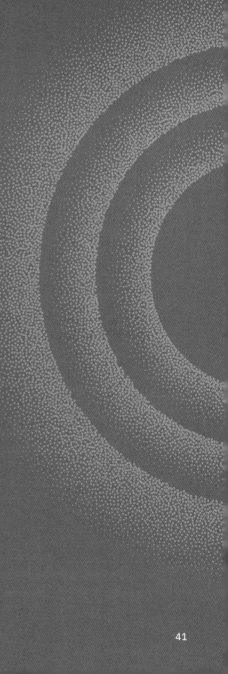

locations have reported turning the music up at the end of the day specifically to disrupt productivity and sweep out the last deadbeat customers—the audio version of flickering the lights at last call.

Beyond conventionally defined music, ambient nature sounds, like crashing ocean waves, or the peaceful burble of a running stream, can also serve to drown out distractions. But as we saw in Chapter 1: Relax, these sounds can be calming to the point of sleepiness...so while they might dial down the stress, they're unlikely to induce the productivity you're looking for. There's a reason every massage room in America sounds like you're lying facedown in a river that cuts through a wind-chime farm. No, if your goal is to tune out distractions *and* get work done, what you need are songs.

## TRAVEL TO THE BEAT OF A DIFFERENT DRUM

Songs are different from other distracting sounds in one critical way: They have rhythm. Some people include ambient sounds in a loose definition of music, and that's fine, but songs have a regular cadence, a tempo...something that repeats. These patterns are partly predictable but with intriguing variations—just the sort of thing our meaning-starved, pattern-hungry brain is always on the lookout for. Music that our brains find meaningful starts with a beat.

"We know that there are rhythms all over nature," Dr. Adam Gazzaley, a neuroscientist and the author of *The Distracted Mind*, told me. "Our brains operate on rhythmic principles, so I think there is a fundamental, deep connection to rhythm. And it's not just how you hear rhythms; it's how you *move* to rhythms."

Since ancient times we've used drums to focus human behavior. Picture the backward-facing drummer keeping multiple lines of straining rowers in sync on a Roman trireme, or the rat-a-tat snare drummer whose cadence holds thousands of marching British redcoats in lockstep. Ancient cultures employed slow drumming to induce trance states, and fast, upbeat rhythms have psychologically prepared warriors for battle throughout history—and still do today. "I think rhythm is in our DNA," suggests world-renowned drummer Jonathan Mover. "It goes all the way back to the very beginning of communication. When one tribe would talk to another or was getting ready to go to war, or celebrating a marriage, it was communicated through drumming. Drumming is really the basis and the root of all of it."

Even in the classic setup of a modern four-piece band, it's the drums that generally produce the steady backbeat everyone else plays to. As neuroscientist Dr. Heather Read, a psychologist and the director of the Brain Computer Interface group, explains: "Neuroscience research finds that temporally predictable beats, like musical rhythms, invoke neural

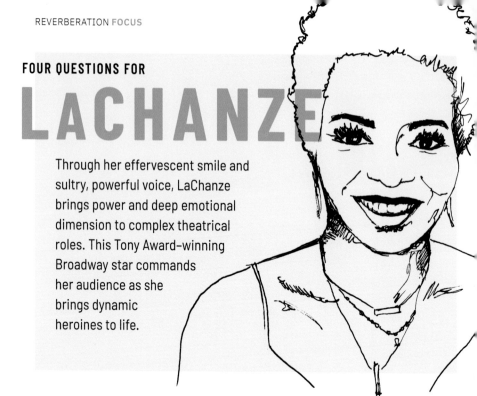

# FOUR QUESTIONS FOR

# LaCHANZE

Through her effervescent smile and sultry, powerful voice, LaChanze brings power and deep emotional dimension to complex theatrical roles. This Tony Award–winning Broadway star commands her audience as she brings dynamic heroines to life.

**Could you describe the different challenges of just singing a song versus singing a song in character?**

When I sing in a concert environment, I'm really open to the response of the audience. So they're right there with me. I'm singing to them. It's like I'm speaking to you right now, we're having an exchange. When I am performing a song within a character, I'm expressing my inner thoughts or emotion, and I'm not necessarily looking to engage with you. It's not necessarily so much of a song onstage as it is communicating the thought or the idea or the emotion of what I'm saying.

So you could say in a sentence, "Oh, what a lovely day." And then if you sing, "What a lovely day," you might have a different opinion of it, depending on how you're feeling. "What a lovely day," can be like, "Well, this really is a f—ed-up day." You can have so many different interpretations of what a lovely day is when you're singing it within a character, as opposed to just singing "A lovely day."

**Is it emotionally exhausting to perform the same part onstage every day?**

You're not really doing the same thing every time, because we're human beings. We wake up one day, we feel amazing. Wake up the next day, we don't feel so hot, or maybe we're vocally tired. It could be a variety of challenges. But since the contract is eight shows a week, you gotta show up regardless of how you feel.

If I get on that stage and I'm playing, let's say Celie from *The Color Purple*, that's already a very draining and heavy story trajectory that I'm on. And, for instance, I may be vocally tired, or it's my eighth performance of the week. And vocally I'm just spent, and emotionally I'm drained because I'm raising two kids, and everything is happening. So I use that. I apply that to the character. I don't try to work as if I'm not tired, I just use it.

It's live theater. The reason people love it is because it's live and there's no hiding behind a cut or retake or edits. You have to be authentically in the moment. Why we love live theater is because we are human beings and we are experiencing this change of emotion, this interactive experience together. Perhaps something makes me emotional in a moment and I'm singing a song and it brings me to tears and I'm swallowing and I'm trying to sing and I can't get it through; the audience is there, they're in the palm of my hand and they're feeling what I'm feeling. It gives people a chance to tap into feelings and allow themselves to experience emotions that they may not have the courage to do in their day-to-day lives.

**Do you think music specifically does that?**

Yes. There's something about tone that reverberates with us spiritually. There's something about a certain pitch, or a certain tone, or how you approach it. One of the things I've been told most of my career is that when I sing, people just feel so much. They feel the joy or something about my voice just makes them feel something, and I just think that it's the reverberation of the sound. Maybe my sound connects with someone spiritually. There are certain singers that I connect with spiritually because I love their sound.

**Do you think music can help get us through grief?**

Absolutely. When I lost someone very important to me in my life, I wanted to just bury my head under the pillow, stay in my bedroom and all that. And I did for some part of it. Depending on what I needed to feel, if I wanted to allow myself to grieve even more, I put on some music that would help me open myself emotionally—music does that. No matter how you're feeling, it just opens your spirit, your mind, and allows you to experience emotion.

Or like if you're in the car and you're on the radio and something comes on and you're like, "Oh, I can't hear that right now." The reason you can't hear it, it's not because you don't love it or don't like the song, it's because you don't wanna experience that emotion in the moment. That's the power of it.

## How do I get that song out of my head?

*That stuck song is popularly known as an* **earworm**. *"Our ability to imagine music, known as musical imagery, is very good," says Dr. Kelly Jakubowski. "Studies have shown we are able to imagine music highly accurately in terms of its pitch, tempo, and volume. An earworm is a specific type of musical imagery experience, in which a piece of music (often just a short section) comes to mind spontaneously and repeats on a loop." The brain doesn't differentiate between earworms and actual musical experiences, and it may call them into service to help regulate moods, just like real music does. For example, subjects have reported hearing the same earworm every time they're stressed. If you're plagued by an earworm, try pulling up the song and listening to it all the way through, or playing a memorable tune you don't associate with a mood (one fan favorite is "God Save the Queen").*

firing patterns that are synchronized to the beat, allowing the brain and body to anticipate and execute upcoming movement." This focuses the band, but it also focuses the listeners: Drums set the tempo, which induces the listeners' brains to produce the specific waves that in turn guide them toward the desired emotional response.

"Rhythms are powerful because their regularity enables prediction of the future," says Dr. Laurel Trainor, a cognitive psychologist and professor at McMaster University. "If someone claps three regular beats in succession, for example, your brain knows exactly where the next beat should occur. There is much evidence to think of the brain as an organ whose function is to continually predict the future, compare its prediction to what happens, and learn from wrong predictions by updating its models of how the world works." In other words, our brains have evolved to always be on the lookout for patterns and rhythms because they make prediction possible—and survival easier. Music opens that door.

What kind of rhythm sets the proper mindset for focus and concentration? You want to guide your brain into beta wave production, reflecting activity in the

Is Lady Gaga's "Bad Romance" stuck in your head? Chewing gum could knock it right out—it's been found that the chewing helps to interfere with our brain's annoying "inner speech."

range of 12–35 Hz. Tempo is important: Up to a point, increased tempo yields increased beta wave production, with faster-paced, upbeat songs between 100 and 180 bpm being a particular sweet spot. (The average pop song is 116 bpm, according to *The Washington Post*.) This insight might explain why, in one study of homework playlists, Olivia Rodrigo's "Drivers License," at 144 bpm, was the most popular of 100,000 songs surveyed. Playing songs you enjoy, according to another study, amplifies beta production even further.

However, familiarity is a double-edged sword: Songs you know and love, as anyone who's ever tried to sing the roof off the car on a road trip knows, can be incredibly distracting. If you want your teenagers to drive safely, for example, your best bet is to forbid them to listen to their own music. In one test of 85 young drivers, teenagers allowed to listen to their choice of music were more easily distracted when driving, made more mistakes, and drove more aggressively. Giving them unfamiliar music improved their scores. The study's author surmised that this was in part because listening to your own music (typically at teen-friendly high volumes) shifts you from the appropriate "shared space" mindset of driving into a more personal-space mindset, in which you forget that your actions can impact others—in this case, literally.

# THE MYTH OF MULTI-TASKING

The philosophy of multitasking holds that your clever brain can perform multiple tasks in parallel. But it can't, because your working memory capacity is absolute. When you *think* you're multitasking—say, watching a documentary while simultaneously responding to emails and talking on the phone—you're really *serial tasking*, shuttling quickly from task to task, giving essentially your full attention to each in order, like a chess grandmaster playing against nine students "at once."

To prove to yourself that multitasking is an illusion, try this simple test from Denmark's Potential Project.

**STEP ONE Perform this serial task.**

Starting with a pen or pencil and a sheet of lined paper, time yourself on this exercise: Write "I am a great multitasker" on the first line, then write the numbers one to 20 on the second line. See how you did.

**STEP TWO Repeat, this time as a multitasker.**

Now time yourself repeating the same task on the third and fourth lines, only this time go back and forth between the lines—i.e., multitasking. First write the letter "I" on the top line, then the number "1" on the bottom line, then "A" on the top line, and "2" on the bottom, and so on, until you've completed the task.

**STEP THREE Wallow in your failure.**

Did it take you twice as long? Four times as long? Could you feel your brain struggling to continually reset from one task to the other?

# THE FLIP SIDE OF FOCUS: MULTITASKING

That brings us to multitasking: everybody's dream, but nobody's reality. (See "The Myth of Multitasking," left.) Music can help us be more productive, whatever we're trying to do, but the way our memory works puts a hard upper limit on the amount we can manage at any one time.

For example: If I tell you a random seven-digit number right now, there's zero chance you'll remember it in a month. But if the cute stranger at the bar whispers their phone number to you, you can typically remember it just long enough to scramble for some way to write it down. This is your **working memory**, your brain's system for holding on to information briefly while deciding whether or not (and how) to store it. Sometimes called the "sticky notes for the brain," working memory utilizes a few different schemes, like visual information (e.g., picturing the numbers on a keypad) and verbal information (e.g., repeating them to yourself), to hold the items in question for about 10 to 15 seconds, before storing or trashing them.

It's one of humanity's greatest party tricks, but our working memory, tragically, has a maximum capacity. The average person can remember just seven or eight things in order—grocery items to pick up, driving directions, etc.—without forgetting individual items. Why seven or eight? The reason is kind of fascinating. To remember a set of things in order, your brain organizes a cluster of **neurons** representing each individual item. It then creates a sequence that expresses those clusters in order, beginning by suppressing the firing of all clusters except the first one, then suppressing all but the second one, and so on. This gets exponentially more difficult the longer the sequence gets—it's estimated to be about 15 times harder to remember seven items than to remember three items. And *that's* why you forgot to pick up the broccoli.

Interestingly, some people on the autism spectrum can effortlessly forge much stronger pathways and remember strings of 100 or more digits—a powerful statement of the value of **neurodiversity**. (Neurodiversity advocates consider atypical mental conditions like autism to be valuable alternative perspectives to be celebrated rather than abnormalities to be corrected.)

The real reason multitasking doesn't work is because switching among multiple mental tasks requires a continuous rebooting of your working memory, starting the process over again every time you switch, just to retain the same information. It makes you feel frantic and busy, when in fact it's severely impairing your productivity—by as much as 40%. According to one study, participants trying to multitask their way through cognitive tests experienced declines in apparent IQ similar to the effects of smoking marijuana or staying up all night.

## MUSIC TO STUDY BY

The pressure to concentrate and retain new information through applied focus is felt acutely by students. Abuse of Adderall, the cheap, addictive "study drug," is now considered an epidemic, with dedicated clinics and treatment centers throughout the country. However, according to American Addiction Centers, unless the user has a clinically diagnosed attention disorder Adderall doesn't actually improve cognitive functioning or the ability to learn. These "mental steroids" may enable short-term performance and serve the immediate goals

**FUN FACT**

For better focus at work, try listening to video game soundtracks, engineered as background music for puzzle-solving and other concentration-heavy tasks.

FOUR QUESTIONS FOR

# KOOL

The co-founding member of Kool & the Gang is behind some of the grooviest jukebox anthems of all time, like "Celebration," "Get Down On It," and "Jungle Boogie." The band is one of the most sampled acts in music and has sold more than 80 million albums.

**How does music help start your day?**

I'll listen to classical, the kind of music that relaxes you, takes the stress away a little bit. It relaxes me. It prepares me for the day—whatever business calls I gotta make or whatever I'm gonna be doing that day. I'm not a TV watcher that much...but in the morning, I gotta think, so I gotta have thinking music, relaxing music, you know, jazz or classical.

**A good pop song has a stickiness to it—once you hear it, it's in your head for the rest of the day. Isn't that the holy grail?**

Well, in some songs, I think it was the simplicity. One of the things that my mother used to tell us was, "You know, everybody likes a good melody, a song that people always can remember." You know, simple songs, and that's how "Funky Stuff" came up, it had a simple melody, very simple, not complicated. "Hollywood Swinging," simplicity.

**Do you find that musicians sometimes overcomplicate songs?**

"Celebration" came after "Ladies' Night," and we were celebrating because we'd turned our careers around. So my brother played that track, and it had a sort of down-home feeling, like down there in Birmingham, Alabama, sipping on some Kool-Aid and you're sitting in a rocking chair. And who knew that that song was going to become what it has become. It wasn't a very complicated song. It wasn't a song with a whole lot of lyrics. It was just how it came together.

You look at "Get Down on It" and [sings melody] *"Get down on it...get down on it..."* There's nothing complicated about that. But it's the way that is put together. My brother started listening to Bob Marley. And that's when he came up with "Get Down on It." Now, it's not reggae, but it had that feel to it.

Yeah, we stayed the course. Sometimes songs can be overcomplicated and kind of miss it. They go too far out there; I kept it simple, basic, like building the house, you start with the drum, you start with the bass and keyboards and the guitar. We just stay to the groove, I don't have to play all that funky stuff, all that hand-popping stuff. And we have had big hits by not doing it. It's the simple titles, it's the beat, it's the groove, you know?

With the Beatles with "Eleanor Rigby," the Beatles' music wasn't really that complicated. It was just how they put it all together. And then "I Wanna Hold Your Hand," that's it.

**What's the happiest song you play?**

In the early days, we didn't have a lead singer. We didn't follow a lead singer, we were just grooving...but we had to start making room for the singer. We like "Hollywood Swinging" and "Jungle Boogie," because our trumpet player, he gets off, you know, he used to play with some rock. We'd be like, "What is he doing?" But this is freedom, when there's no singer in the way. I've got nothing against singers, but it's all about the musicianship, and we are just grooving and growing...Wow. It's simplicity. Okay. Simplicity.

## Does background music help with studying?

*Not particularly. Music can be terribly distracting, especially music with lyrics, and especially the favorite songs the typical student is likely to put on an all-night-cram playlist. Listening to music, even in the background, requires a little bit of attention, which reduces your working memory capacity and lowers reading comprehension. However, music is great for motivation and setting a productive mood, meaning your best plan is likely to study either in silence or with instrumental music at a low volume. Your playlist is probably too distractingly incredible to help you focus, but by all means play your favorite music before you study, on study breaks, and as a treat after you're done, so you get the motivational benefits without the distraction.*

of the time-strapped student, but they don't do the brain any favors.

Music offers a low-risk, zero-cost alternative—when used wisely. Classical music in particular has been shown to reduce the kind of stress students are under and improve their mood, even if the so-called "Mozart effect," in which classical music is thought to improve your general intelligence, is widely considered to be bunk. (The idea spread from a popular misreading of a study that found only modestly improved spatial-reasoning skills in a small set of college students.) But music has been shown to help students excel temporarily, likely thanks to its proven ability to reduce stress, improve mood, and provide chemical rewards. It doesn't make you smarter—it just makes you a happier student, and sometimes that helps.

Ambient electronic music, which is also low on distracting lyrics, can provide similar benefits, prepping student brains to more easily absorb and interpret new information. One small study of musical genres declared electronic dance music, or EDM, the most helpful genre tested, increasing speed and accuracy for nine of 10 participants across a range of tasks, including mathematical word problems, spell-checking, and abstract reasoning. Steve Aoki has this to say about EDM music: "The music opens up another portal that allows for more ingestion or more absorption. At the end of the day, what it's all about is to optimize my absorption rate of whatever I'm doing."

Another lyric-free alternative is the musical scores from video games, which are often specifically designed to elicit emotional reactions while keeping your brain focused on evaluating, navigating, and completing tasks.

That's right, killing wave after wave of zombies is a task—an ugly, difficult, never-ending task. Which brings us to one of the most interesting areas of focus: the **flow state**, otherwise known as "the zone."

## EYE OF THE TIGER

Have you ever been so thoroughly immersed in an activity—running, writing, cooking—that you lost all sense of time passing? This is colloquially known as being "in the zone," and it's a much desired—but often elusive—state of hyperfocus. Challenges simplify and distractions fall away; you feel confident, motivated, and clear on what needs to be done. And when you look up from what you're doing, you may be surprised to find that hours have flown by.

"So if you're lonely, you know I'm here waiting for you

I'm just a cross-hair, I'm just a shot away from you."

FRANZ FERDINAND "Take Me Out"

Athletes in the zone, to take one example, report experiencing unusual clarity, quicker reaction time, and sharpened sensory perception. For basketball players, the rim grows larger; for batters, the incoming fastball looks as big and easy to hit as a grapefruit. "No longer conscious of my movement, I discovered a new unity with nature," said Roger Bannister, trying to describe his mental state when he ran the world's first four-minute mile.

Scientists call this condition a "flow state," and nobody's more familiar with flow states than songwriters and performers, who often rely on that level of productive hyper-focus to create their best work. Rock legend Mick Fleetwood describes it like this: "I get into a moment where I feel completely safe, and I know that I'm in the zone, and it's timeless and magical." And Steve Aoki says, "Once you're in flow, things just come naturally—you don't have to think twice about it. My brain cycles during that time, when I'm writing and composing this kind of music, are supercharged. And when you're in that state of flow, you just need to stay there as long as you can, because it's hard to get there."

You can leverage music to help guide you into a productive flow state, and you don't have to be a professional musician to do it. "I recommend 'pre-gaming' with music to boost your mood and make you that much more primed to get into the groove," says Diana Saville, cofounder and CEO of brain-industry accelerator BrainMind. "I find it easier to 'lose myself' in creative pursuits when I mindfully prepare the setting. Since listening to preferred music can improve your mood (this has been demonstrated in numerous studies, but we all know this to be true anecdotally), I recommend listening to curated playlists both before and during the activity in which you hope to achieve flow."

So before your activity, invest the time to first create a playlist of favorite songs—you'll get the "lost" time back in increased productivity once you get your flow on.

## CONCLUSION

The dream is to be able to leverage the power of focus at will and turn every crisis moment into an opportunity for superhuman performance...or at least high efficiency and effectiveness. To dispel distraction and crush the exam, get through a week's worth of work in a day, summon breakthrough ideas by sheer force of will.

We're not quite there yet—there's no magic plan for summoning flow states on command. But we know, now, that music can be a powerful force to help us trigger a more productive state of mind.

Applying music to the world of work has already led to some major successes. Workers allowed to listen to music complete tasks faster and generate better ideas than their music-deprived counterparts. Surgeons who listen to music in preparation for surgery perform operations more efficiently and accurately. Athletes work out harder and longer, and achieve better results on the field, when they listen to music. The list goes on and on. We've just begun to rock the world. ∎

# TAKEAWAYS

*Music works on a number of levels to help us overcome the limitations of our wandering minds and stay focused. It fills the background and drowns out conversation, its cadence can induce the production of the brain waves we need to stay on point, and its familiarity or novelty can alter our mindset through mirroring effects. Here are some ways to make this real in your life.*

**TO WAKE UP YOUR TEAM BEFORE THE BIG PRESENTATION** Playing a "wake up and rally" song at 100–180 bpm—like "Mr. Brightside" by The Killers at 148 bpm or "9 to 5" by Dolly Parton at 102 bpm (there are widgets online to find the bpm of any song)—can be enough to generate the beta waves you need for high focus.

**TO STAY PRODUCTIVE WHILE WORKING FROM HOME** Occupy and distract your wandering brain with a background track. Any ambient sound is helpful, but music is even better. You want familiar songs, played at a moderate sound level, but not your favorites— this way you can build a productive personal space bubble without being distracted from the work at hand.

**TO STUDY WELL AND REMEMBER WHAT YOU'RE STUDYING** Try EDM or light classical music to provide the stress reduction and mood-boosting value of music without giving your mind lyrics to subconsciously process, especially if your studies involve reading, writing, or critical thinking.

**TO STAY FOCUSED WHILE DRIVING** Play something unfamiliar to open up your "personal space" and keep yourself from getting so comfortable you subconsciously forget you're traveling at a dangerous speed in a moving vehicle. Again, you're looking for something in the 100–180 bpm range, like "Up" by Cardi B at 166 bpm or "Semi-Charmed Life" by Third Eye Blind at 102 bpm, to keep you focused. (Of course, if you're not so much bored as tired, pull over safely!)

# HELLO,

## IS IT ME

**LIONEL RICHIE**
"Hello"

# YOU'RE LOOKING FOR?

**YOU'LL NEVER FORGET THE FEELING:** the spring in your step as you float back to the car after that perfect date where you first realized that this one's...special. This isn't just a crush, clearly, and it isn't just lust, though God knows there's some of that, too. No, this is something animal and real, something you feel in your bones. And as cliché as it sounds, you're already counting down the minutes until you can see this not-like-anybody-else creature again. You're in the driver's seat, now, and your heart is racing like Usain Bolt. It's hard to focus on anything else with your head spinning like this: You're in love, of course.

Or maybe you have food poisoning.

But let's assume you're in love. What you may not realize is that your brain has already started undergoing a chemical transformation—one that's been hardwired into human DNA for thousands of years. If it feels like your thoughts are swimming, it's because they are: Your brain is surging with that sweet "happy hormone" dopamine and its warm 'n' cuddly cousin, **oxytocin**. At the same time, your level of **serotonin**—the neurotransmitter that affects your sense of being in control—is dropping discreetly.

Most of the action takes place in your ventral tegmental area and caudate nucleus, brain regions responsible for pleasure, motivation, and processing rewards. But the prefrontal cortex—our brain's command-and-control center for reasoning—also drops into low gear when we're in love, and the amygdalae, key components of our threat-response system, also rev down. You experience all of this as giddiness, warmth, obsessive thoughts, a pinch of instability and recklessness, and a general appetite for crazy.

Falling in love feels good and makes you obsessed because evolution wants babies. When you find a compatible partner, your brain—using tools carefully honed over millennia of natural selection—seizes the opportunity, rewarding you with a constant drip of feel-good **hormones** designed to motivate you to stick out the relationship, all the way to the diapered conclusion evolution wants.

But what's really remarkable about the chemistry of falling in love is the similarities it shares with the chemistry of listening to music. Music can stimulate the release of the same neurotransmitters, and light up all the same brain pathways, as romantic love—your brain's reward system, by and large, seems to respond similarly to both stimuli. When

you say you love a song, it turns out, you really mean it.

Is it any wonder that more than half of all songs are about love? It's been estimated that humans have recorded 100 million different love songs, up to and including half of this week's Billboard 100. (The very first human voice recording, Alexander Graham Bell's plaintive "Mr. Watson, come here. I want you...," even sounds like the start of a love song.) Clearly, we love singing about love. But why this curious obsession with declaring our love through song? And how can we leverage music's power to improve romance, sex, commitment, and all that good stuff in between?

Buckle up—this is going to be a bumpy ride.

## YOU LIGHT UP MY VENTRAL TEGMENTAL AREA

Reproduction is every living creature's Job One, and your brain takes the chemistry of love and lust very seriously. Just showing people photos of their romantic partners lights up their brain's reward centers; thinking about the one we love can provide a narcotic-like high. At the same time, the brain in love experiences an increase in the stress hormone **norepinephrine**, which increases heart rate and blood pressure, producing more of what you might call a methamphetamine kick. These chemical rewards, like

Almost one out of five people admitted to saying "I love you" as a result of a song they heard together. **Be careful out there.**

*all* chemical rewards, in turn increase our motivation to pursue and acquire more rewards. Talk about being addicted to love.

Love and lust appear to induce separate but overlapping neural responses in the brain: They both produce a "high," are addictive, and affect many of the same brain regions. But they are distinct enough that you can all too easily find yourself in the uncomfortable position of being in love with one person and in lust with another. In fact, it's even more complicated than that, according to Dr. Helen Fisher, senior research fellow at the Kinsey Institute and chief science advisor to Match.com. "We've evolved three distinctly different brain systems for mating and reproduction. One is the sex drive, the second is romantic love, and the third is feeling a deep attachment," she told me. "They're different brain systems, just like fear and anger and surprise are different brain systems." Think of romantic love as the sweet viral proposal video, attachment as the old couple holding hands on the park bench, and lust as...well, as the annoying couple in the hotel room right above yours. It's fantastic when all of these separate brain systems cooperate in one relationship, but it doesn't always work out that way.

It may seem self-evident that love is an emotion, but it's actually something much more profound—an ancient survival mechanism that we share with other animals, from foxes to birds to elephants. "Romantic love is a profoundly basic drive," says Dr. Fisher. "It evolved millions of years ago to enable us to focus our mating energy on a particular individual." The ventral tegmental area, that factory that creates our **feelings** of romantic love, lies right next to the factory that creates feelings of hunger and thirst. "Hunger is there to keep you alive today," says Dr. Fisher, "and romantic love drives you to form a partnership and send your DNA into tomorrow."

Over a long-term romantic relationship, our real-time love, lust, and attachment responses are mediated by structural brain changes: For example, the brains of long-haul couples show increased activity in the ventral pallidum, a region of the brain rich with oxytocin and vasopressin receptors that facilitate long-term pair bonding and attachment. And that's how your long-term commitment becomes structurally fixed, keeping you from being tempted by Susie from accounts payable, and improving your health into the bargain. Notes Dr. Fisher: "People who are in good long-term partnerships tend to live five to seven years longer."

Brains reward bodies for engaging in activities that create more brains: This is the all-natural evolutionary system that's worked well for thousands of years. But music can easily be dragooned into love's sweet service to step up the pace. Whether you're trying to encourage romance, level up your bedroom game, make love stay, or soothe a broken heart when love flits off, you can leverage music to put your brain (and sometimes the brains of others) into a mindset that optimizes your chances for success.

**FUN FACT**

# Want to make your lovemaking more rhythmic? Almost half of people polled in one survey said the right music does the trick.

# MEET YOUR NEUROTRANSMITTERS

*A few of the important chemicals coursing through your brain, and what they do*

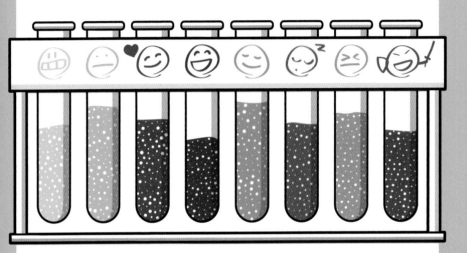

**GLUTAMATE** The most common neurochemical is linked to learning and memory. Too much can lead to impulsive/violent behavior.

**GABA** If glutamate is the throttle, GABA is the brake; it increases tranquility and makes sure you don't get too aggressive.

**OXYTOCIN** This "love drug" makes you fall in love and feel connected. It also helps nursing mothers bond with their babies.

**DOPAMINE** The "reward" chemical associated with pleasure of many kinds. Too little can lead to depression; too much can lead to addiction.

**SEROTONIN** Modulates mood; associated with general serenity. Low levels can lead to depression.

**ADENOSINE** This neurotransmitter builds up throughout the day; when enough has accumulated, you will feel sleepy.

**ENDORPHINS** These are your body's painkillers. At high levels you get relaxed and euphoric; at low levels you may be more sensitive to pain.

**NORADRENALINE** Like its cousin adrenaline, norepinephrine heightens energy, and is involved with the fight-or-flight response and with elevating heart rate.

FOUR QUESTIONS FOR

# MICK FLEETWOOD

The irrepressible co-founding member of rock legends Fleetwood Mac is often regarded as the glue that kept this famously volatile band together. The charismatic drummer has also recently added "viral TikTok star" to a long list of accomplishments.

**Is it okay to be in silence?**

Whenever there's a barren moment emotionally, I can't stand not to feel. I will get triggered by music and I will always go there. I probably involuntarily do it way more than I'm confessing to. I can be sitting on a beach. I'm like, "Am I a drummer?" I'm just Mick. Sometimes when I feel a bit flat, I will absolutely retreat, and something will trigger me. And it's almost always music.

**When writing a song, is it about you, or the listener?**

A lot of what I look to in music is a reminder of how vulnerable we all are, and from that, you've gone up huge amounts of strength, whether you're really aware of that or not. Talking for myself, I think that vulnerability is a blessing and that that's how I listen. It's a reminder of something which is so unbelievably effectual in the art form known as harmony and music and the continuity of rhythm. When it comes together like an alchemist, if your little tentacles are at least halfway open, there is nothing like it.

"Dreams" is a reflective, dramatic, deeply personal song. And fans found it exceptionally relatable.

Fleetwood Mac, right from the beginning...came from hanging onto everything about what was the fascination of feeling. We're not slick. There was a drama to that, the chemistry of Lindsey, Stevie, and let's just say the band. It became a backdrop that was so personal and very real.

Maybe we were naive and we revealed too much, but really looking back on it, that was what was really important about it. So when you say that it was sort of bringing down and identifying with hundreds of millions of people in a certain way, that's how powerful it is. God knows we didn't think about that. It was just really a diary of feelings. And later on, it becomes more of a sort of soap opera for us. I know the soap opera actually became almost more of a burden as you realize...I'd never looked at it like that, but "Dreams" plugged into a lot of the available people who identified very specifically with a certain type of feeling. Again, that's the power of what we're talking about.

**As a drummer, how do you approach creating a "sad song" differently than a "happy song"?**

Listen, like crazy. Become the biggest sponge known to mankind and soak it up, which was my training and was also part of the beginning of the conversation. Listening to music was an escape. And in the escape I learned how important that journey is and how much you can put into supporting your version. Always try to become part of it. That's what the magic is. You don't always get it, but we know what that is. There are structures that have a cause and effect that are literally chemically connected to very, very powerful emotions.

I'm the king sucker, looking for that. So I'm an addict really.

**What kind of physical lift do you get from playing the drums?**

You can't get away from the fact that everything has a rhythm to it, in its most simple form. It's that powerful, like sitting in front of a wave...

There's a plateau that I always look forward to. I get incredibly wound up and think, "I can't even do a drum roll before I go on-stage," which is like a child. And then I'm in the playground. Then there are no bullies. But then I get into a full moment where I feel completely safe, and I know that I'm in the zone and it's timeless and magical. Almost always.

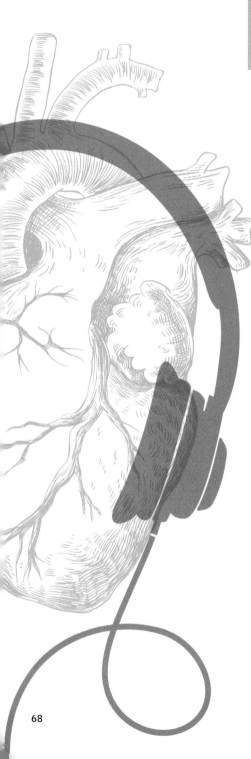

## HOW MUSIC HELPS

Music is very efficient at recruiting the emotional centers of the brain and activating physical responses. Dr. Francesca Dillman Carpentier, media psychologist and professor at the University of North Carolina at Chapel Hill, explained it to me like this: "Sounds can affect us physiologically, as well as emotionally. Certain high-pitched notes can make our skin feel chills. Certain low-pitched sounds make our chest physically vibrate, which we can experience as pleasurable or even misattribute as an emotion stirring our heart." This is why the right song can give you goosebumps or make you cry. That said, music (like love) isn't just about the reptilian emotional response. "Biologically, music stimulates the **limbic systems** of the brain, these fundamentally primitive systems that deal with survival and arousal and fear and sex and all of these things that are really basic to our existence," says Dr. Charles Limb, surgeon, neuroscientist, and musician at the University of California, San Francisco. "But music isn't as much brainstem as it is higher-order thought."

When you're trying to inflame someone's passion, get those upbeat love songs going. It's been shown that listening to positive music when interacting with whomever you have your eye on can help them view you more positively. But that's just the beginning: Humans are prone to subconsciously construing a song's lyrics to be directed specifically at them. Why? "We are evolutionarily programmed to respond to particular cues in the human voice and to perceive them as expressing particular emotions," musician and PhD Sandra Garrido of Western Sydney University told ABC News. "And when those same features occur in music, we respond to that in the same way, as if it was a person in front of us doing that." So if you're preparing to make your move on the girl of your dreams, you can be sporting your least-stained button-down and your most reliable cologne, but if the background music is Puddle of Mudd's whiny "She Hates Me," you're fighting an uphill battle.

In fact, we let songs speak for us when we can't quite do it ourselves, don't we? We quote significant lyrics in our love notes and anniversary cards, we dedicate songs to one another at wedding receptions, and man, oh man, do we create romantic playlists (see sidebar, page 78). Tellingly, this tends to happen more in the early stages of a relationship, when we're still unsure of ourselves and trying to establish mutual trust. "A special song can provide us with perhaps the only tool we have to communicate what we feel to a significant other," Carpentier told me. "Song lyrics can represent for us words we can't imagine ourselves but that capture our feelings, or words we might feel too silly, too uncomfortable, or too afraid to say."

Memory is another potential focal point for romantic leverage: Sharing a familiarity with a particular piece of music can be extremely important for emotional engagement between partners. Once you associate a song with a specific emotional experience—say, whatever was playing in the background for you and your boyfriend's first Naked Cupcakes Wednesday—that association is likely to stay with the song every time you hear it. Replaying the songs that soundtracked your best moments can help strengthen

those emotional memories and provide some microdoses of passion throughout a long-term relationship.

Scientifically valid movie reference: In the classic movie *Casablanca*, Humphrey Bogart's nightclub owner has ordered his piano player, Sam, never to play the song "As Time Goes By," because it brings back emotion-rich memories of a wonderful time he spent with a woman who ultimately left him—and when she resurfaces and requests the song, it pulls them right back into each other's orbit.

## THIS MAGIC MOMENT

The classic first-date question "So, what kind of music do you listen to?" might sound like a throwaway conversation starter—something to kick into the circle only after you've run out of ways to talk about the weather or avoid politics. In fact, music preferences can be quite revealing. "I think music does communicate things that are very personal, that are very direct, that are immediate, and that don't necessarily have to be verbal," says Tod Machover, composer, professor, and inventor at the MIT Media Lab. When we use music to speak for us, we tell what can be a very interesting story.

In fact, it may be one of the greatest compatibility questions a date could ask. Music is a key part of our identity and is so fundamental to our idea of who we are that it can change the course of relationships. A recent survey of audiences conducted by the ticket seller TickPick suggests that a couple has only a 2%

**59% of people in one study said they find a possible partner more attractive if that partner is playing music they like.**

chance of success—2%!—if their musical tastes are in opposition. You can be a little bit country if they're a little bit rock & roll, but if they *hate* your country music, honey, you got to two-step your way on out the door.

Why can couples survive having different religious beliefs and different ethnic backgrounds, but not different musical preferences? Because our choice of music telegraphs important information about our value system. Diana Boer, the leader of an eye-opening study on shared musical preferences in relationships, summarized it like this: "Individuals who reject conservative values and who endorse openness to change values like listening to rock and punk; individuals who are guided by self-enhancing and openness values tend to like popular music, such as international pop and hip-hop; and individuals with self-transcendent value priorities like listening to jazz and classical. These associations seem to hold across Western and non-Western cultures."

We gravitate toward certain people not because they enjoy the same music we do, but because their favorite music is a proxy for the kind of person they are. (In Chapter 9: Become, we'll get into what this means for individuals.) Consider the phenomenon of "our song," a couple's signature tune that sparks a fond exchange of googly eyes every time it plays. For the 60% of serious couples who have one, "our song" is really a badge of compatibility—significant because it's a reminder of the moment earlier in their relationship when they first realized they shared similar values.

It's probably no surprise that couples who have an "our song" report higher levels of intimacy than those who don't. Generally, couples who enjoy music together experience improved communication, emotional connection, and overall relationship satisfaction; they spend more

time with loved ones; they hug more. The initial bonding from shared music—and continual reinforcement of that bond through shared musical memories—seems to make falling in love and staying in love easier. Not to mention *making love*. Which brings us to...

## LET'S GET IT ON

It's no secret that a little music in the bedroom can help set the mood. But that's just the beginning—the right songs can actually help with virtually every part of the ancient and mystical ritual of doing the nasty.

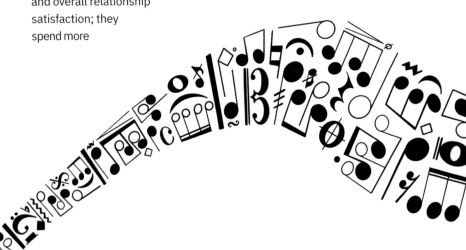

Let's start with stress: Few things can slam the brakes on a potentially hot and heavy evening like unmanaged tension. But as we know from Chapter 1: Relax, music can be a profound, low-cost, and only mildly addictive stress reliever. Slow music, clocking in around 60 bpm (like "Sign of the Times" by Harry Styles or "Tenderly" by Etta James), can stimulate the alpha brain waves that induce feelings of relaxation in you as well as in your partner.

If stress isn't the issue, and you want to make sure nobody gets too tired to tango, a little upbeat music in that 80–100 bpm range, like "Beast of Burden" by The Rolling Stones or "Crazy in Love" by Beyoncé, will entrain all the brains in the room to a higher activity level; the busy beta state's power to keep your mind from wandering can alleviate sexual anxiety, performance fears, negative body-image thoughts, or anything else that's potentially distracting.

Music of any kind is a good neurological warm-up to getting down and dirty. Your brain doesn't really discriminate between sex and other pleasant experiences, such as eating a delicious meal or acing your driving test; it all lights up similar pathways and dispenses the same chemical rewards. And that means listening to music you enjoy before sex is analogous to foreplay for your brain. A solid playlist (see sidebar, page 78) can ramp up your body's natural chemical high, increasing arousal in the brain's "pleasure zones" even before your socks hit the floor.

FOUR QUESTIONS FOR

# STEVE AOKI

Global star deejay, producer, EDM superstar, and brain-health enthusiast Steve Aoki is a bona fide force of nature. Steve is animated by parallel passions: producing, fashion design, NFTs, and his brain-science-focused Aoki Foundation.

**How do you use music during your day?**

Music definitely has a functional role for me. I wouldn't say I'm a passive listener...I use it as a tool, in some ways it's an emotional tool...It brought me closer to myself. It was my security blanket. It was my community. It was kind of my purpose of life for when I was a kid, which is why I picked up instruments and why I dressed the way I did, chose the friends the way I chose, and also directed my path in life, future career, all those decisions were because music touched me emotionally in a way that nothing else could. It's very much like that feeling of love when love takes over.

When I think about my daily activities, when I meditate, I listen to this very particular high-frequency kind of music, so I can get to the optimal stage of meditation. Whether it's a placebo effect or not, I go for that. I try to think more scientifically in nature about my process, why I choose to do the things I do. I try to be as efficient as possible with my time management and time investment and the things that I do. When I work out, I listen to certain music so it pushes me to that point of drive, where I can really have the most efficient activity. When I'm doing different activities, like if I'm in the kitchen, I'm listening to a different kind of music. There's different purposes of music for different reasons. And then when I'm in the studio, I have a completely different way of processing music as well.

**Do you feel the difference when you're not "using" music?**

Yes, I do, because I feel naked sometimes when I'm not doing it. I would say I feel more vulnerable, but actually I feel more vulnerable with the music because that's where you want to be in those moments. You want to be more free and open, so maybe I'm more closed off without the music, the music at least opens up another portal...

**Your shows have been described as religious experiences. Is that how you see it?**

The great thing about music is that it translates across all worlds, all industries, all religions, cultures, everything. It is God, it literally is omnipotent. It literally can live everywhere and anywhere all at the same time.

**Is the music you listen to similar to what you perform or are you going completely out of genre?**

A lot of genres for sure, because it's more functional, you know? I'm not listening to bangers I'm dropping at the festival when I'm meditating, I'm taking out all the drums. I want to be floating. I want to be levitating. And in order to be in that kind of mental state, you need to have music that gets you to that place.

## What is it about "sexy" music that makes it work?

*All music can juice the neural pathways associated with plea-sure, but the bedroom can be a complicated place. According to Daniel Müllensiefen, a music psychologist who analyzed 2,000 songs Spotify users claimed to use in the bedroom, it's not the lyrics that make a song sexy: It's unflashy vocals with circular, repetitive melodies. "Anything that is distracting or demands attention, or has elements of the unexpected, is not so good for romance," Müllensiefen told the* Independent. *His analysis discov-ered that people's playlists tend to contain a lot of classic tracks, including songs from Marvin Gaye and Barry White, the* Dirty Dancing *soundtrack, and Maurice Ravel's Boléro, a 17-minute orchestral ballet from 1928 with a churning, repetitive melody that's been help-ing lovers get into a groove for generations.*

"One of the things that happens when people listen to music out loud together is that their neurons fire synchronously with one another," says Daniel Levitin, in an interview with Sonos about their study of the musical habits of 30,000 listeners. "And for reasons we don't completely understand, this releases oxytocin." Is it any wonder that couples that listen to music out loud together have 67% more sex (as the study found) than those who don't?

When it comes to sex itself, music's superpowers are legendary. Listening to music can increase heart rate and breathing, diminish inhibitions and stress, and increase emotional arousal and libido. Music even helpfully covers up unattractive sounds. And don't get me started on the rhythm: Rhythmic music stimulates the **cerebellum**, the part of the brain associated with motor control and movement, which is what makes us want to dance, sway, or play terrible air-guitar to our favorite tunes. A survey that pored over the go-to sex playlists of college students found that they tended to cluster around either 80 bpm or 130 bpm...tempos that corre-spond with people's, um, hip rhythms when they're taking it slow or enjoying a quickie.

Now, after all that, whether music actually enhances pleasure during sex is still an open question. In an interview in *Men's Health*, the neuropsychiatrist Louann Bri-zendine says, "The dopamine rush from sexual experience completely outweighs

"Don't ya hear me talking, baby? Love me now or I'll go crazy."

MARY J. BLIGE "Sweet Thing"

and overwhelms anything that comes from music—it's like a tsunami relative to a small little wave." Clearly, more research is needed...much, much more research.

## BREAKING UP IS HARD TO DO

All good things come to an end, and for many of us, falling in love all too often leads to falling out of love. (Playing Journey's "Don't Stop Believin'" en route to a climax is sometimes a factor.) It's natural to feel lonely after a breakup: Evolution puts a premium on companionship that promises babies, and when that companionship ends, the feel-good tap is turned off. You feel shitty in part because you're not getting your rewards—effectively, you're being denied the high level of dopamine you've grown accustomed to, and forced to go cold turkey. You might even be tempted to consider an ill-advised rebound relationship just to get those tasty neurotransmitters flowing again.

# How to Build a **Sexytime** Playlist

"*Variety is important for any playlist, to avoid getting into an emotional rut,*" *says Carpentier.* "*I feel the formula is a good playlist that gets a bit more intense in the fullness—harmonies and instrumentation—and a bit faster in the tempo as the list plays on.*" *Other tips:*

**Start with music you both like.**

*If you want to build closer social bonds, choose music that celebrates your common tastes.*

**Low and slow wins the race.**

*When people are feeling down and dirty, their voices generally drop into a lower pitch range; music with that low pitch can move the mood in that direction.*

**Include songs where the vocalist is trying to sound seductive.**

*Primates are literal; studies have shown that we're more likely to be romantically stimulated by a song when the artist's voice has a loving tone.*

**Add tunes from early in your relationship.**

*Nostalgia makes us feel protected, removing inhibitions, and a song's specific romantic associations can set the table for bedroom bliss.*

**Start sweet, then build the beat.**

*Romance is the appetizer; sex is the meal. And the dessert? Chocolate-covered strawberries are nice.*

But you don't have to—there's a methadone for that.

A recent study proved what blues fans have always known: Music can help those struggling with sadness feel more connected to people, even when those other people aren't in the room. Researchers showed that when people of various moods listened to music of their own choosing for 20 minutes, it elevated their mood and reduced loneliness; the researchers hypothesized that spending time with a good song was like spending time with an empathetic friend.

Ah, but which songs to choose? This matters, of course. When it comes to songs that help us get over a breakup, the selection can largely be broken down into two categories: songs that help you cry it out, and songs that remind you you're better off without that idiot. Many of us would instinctively reach for something sad and relevant, like Toni Braxton's "Un-Break My Heart." Turns out that's not a bad place to start.

To lift yourself out of heartbreak, it's best to reach for albums that are what psychologists call "mood congruent." Songs that match your mood *get* you; they say what you're feeling, and science says they make you feel understood at a deep level, which is most likely what you need right about now. This can be why certain songs or albums that strike just the right nerve can become cultural juggernauts. As Mick Fleetwood explains, discussing the megahit "Dreams" from Fleetwood Mac's 27-time platinum-selling album *Rumours*: "It was really just a diary of our feelings. But 'Dreams' plugged into a lot of the emotions of people who identified very specifically with a certain type of feeling. That's the power of what we're talking about."

Instead of rushing to find a song to make you happier, finding a song that understands you is a better first step in the healing process, according to experts. "Matching the sound to the mood can provide an outlet for people in a breakup

One less reason to sing the blues for fans of that genre: They clocked an average 16 minutes of performative endurance in bed, the longest of any genre surveyed in one study.

to reattribute their feelings to the song, while letting the lyrics distract them," Carpentier told me. "Then, they can shift to music that's a little more upbeat to start lifting their mood, which can provide some protection when they're ready to face the breakup."

Something fascinating that researchers have discovered is that sad songs don't make people more sad—they make them more *nostalgic*. And that's a good thing, in the context of a breakup, because, as long as you can avoid songs that remind you of *this* particular breakup, nostalgia can be a productive way to start finding your way back. "Build up your sense of identity and listen to music that has personal meaning and has been with you for a long time—that defines who you are," advised Professor Bill Thompson of Macquarie University in *ABC Everyday*. The point is to generally move away from ruminating and wallowing, which can be a necessary starting point but isn't healthy over the long haul, and toward reflecting and reframing. "Ruminating is something people can fall into easily. It's a comfort, because you're used to going over old ground, but it's not an effective strategy for moving on," says Thompson.

A breakup is a journey, but the pace is up to you. "Each person has go-to coping strategies they employ in stages as they eventually work toward healing," says Carpentier. "Those who need space at first to escape their negative feelings and breathe before they start addressing their hurt and moving on might be able to use music that has nothing to do with romance to begin repairing their moods and get that needed breath. For those who are the type that tackle a problem head-on from the beginning, listening to breakup songs that have an empowering vibe can help bring hope while allowing them to process what's happened." ■

# TAKEAWAYS

*Shakespeare called music "the food of love" in* Twelfth Night, *and there is no aspect of love, from the first kiss to the last goodbye, that music can't improve upon or help you navigate. Music supports intimacy, powers long-term bonding, says the things we can't, makes us feel safe and understood, and, when our hearts get broken along the way, even helps us move on.*

**TO GET YOU-KNOW-WHO TO FALL IN LOVE** Soundtrack your get-togethers with positive, familiar, upbeat songs that say the things you aren't quite ready to say. Don't sabotage your romantic playlist with a song about going it alone—that's as ill-advised as showing an in-flight movie with an airplane crash (as they did in the movie *Airplane*).

**TO KEEP A RELATIONSHIP TIGHT** Listen to shared favorites out loud with your partner, singing along when appropriate, and put any songs with special meaning on permanent speed dial. The dopamine, oxytocin, and nostalgia—which will be re-released when the song is played again—will strengthen your bond.

**TO IMPROVE SEX** Listen to music that clocks in at either 80 bpm (for slow lovemaking—e.g., "I Just Want to Make Love to You" by Muddy Waters) or 130 bpm (for banging—e.g., "Next Lifetime" by Erykah Badu) and crank the bass. Songs with circular, repetitive melodies are best for finding your rhythm—in every sense—without throwing you off your game.

**TO RECOVER FROM A BREAKUP** Start with sad but familiar songs that let you put your dejection in perspective while still allowing you some wallowing time. When it starts feeling cheesy, step it up slowly to more upbeat fare to break out of the blues and reorient yourself to a more promising future.

# THRIVE

# AND YOU
# WHEN THE
# TOLD THAT
# WHAT YOU
# YOU CAN
# OLD.

**BILLY JOEL**
"Vienna"

# KNOW THAT TRUTH IS YOU CAN GET WANT OR JUST GET

**YOU ARE DYING.** Even as you read this paragraph, your hair is getting marginally grayer and sparser; your skin is thinning and wrinkling. Sooner or later, the once-long odds of cancer, heart attack, or stroke will begin to swing against you. Your organs will begin to stutter and falter, your mind will start playing evil tricks. Even your bones—destined to outlive you—will grow increasingly brittle, until even a modest midnight-snack-related stumble in the kitchen could put you in a cast.

What a drag it is getting old, as Mick Jagger says...and what a pleasant way to begin a chapter. But the fact is, long before our bodies betray us, we typically betray our bodies. We drink too much, smoke too much, take too many pills, and stuff ourselves to bursting with all the wrong foods. We spend our days in stressful, sedentary jobs and our nights in toxic relationships. We soak up environmental poisons like pesticides, microplastics, heavy metals, and good old smog like there's no tomorrow. And then, one day, there isn't.

It's hard to stay healthy, and it's easy to take your health for granted—until something goes south. For most of us, health isn't something we work on, it's something we check on. We eat and do what we like, then wait passively on the gurney in the exam room for the doctor to return with our results. But what else are we supposed to do? After all, it's not like you can decrease your own pain, lower your own blood pressure, increase your happiness, reduce your stress, or stave off mental and physical decline by sheer force of will.

Or...can you?

Consider for a moment the amazing chemical factory that is your brain and its connected systems. Over here, you produce **endorphins** that block pain. Over here you make sweet, sweet dopamine, the pleasure potion; adrenaline, the ultimate pick-me-up; serotonin to regulate your mood; acetylcholine to improve your memory; gamma globulin to reinforce your immune system; and on and on and on. And it's all made right here on your top floor.

Keep in mind, these drugs aren't mild store-brand substitutes for real-world cocaine, caffeine, nicotine, etc. ...it's quite the other way around. Cocaine, for example, doesn't itself give you a rush—it tricks your brain into dropping a load of dopamine, and *that* gives you the rush. You are always, to a certain extent, getting high on your own supply.

Your brain is a fully automated wellness factory—the trick is learning how to run it. If you could only find ways to reliably push your biological machinery to produce the substances you need for the effects you want, you could deliver your own shots, so to speak, and improve your physical and mental health across a wide range of conditions. Happily, we already know a great deal about what sorts of real-world inputs can coax your brain into producing the fixes you need.

And none of them are more promising than music.

## YOU'VE GOT THE MUSIC IN YOU

I asked Dr. Joy Allen, chair of the Music Therapy Department at Berklee College of Music, the founding and acting director of Berklee's Music and Health Institute, and one of the foremost avatars of the burgeoning **music therapy** movement, to give me a thoughtful definition of music therapy. "Music therapy is an established healthcare profession and complementary treatment modality," she told me. "Music therapists utilize music, and the relationship that develops between the client and the therapist through the music, to promote healing and enhance quality of life. By using music to address health, music therapists essentially recognize the multidimensional aspects of self as well as the need for multidimensional tools to access, explore, re-create, and/or create new ways of being.

"After a thorough assessment of strengths and challenges, the music therapists devise treatment plans that incorporate music-based experiences for targeted health outcomes," she continued. "Music therapists are working in leading medical, educational, and community-based settings to improve health outcomes within a wide variety of medical/developmental conditions, including experiences designed to foster developmental gains in premature infants; promote conflict resolution, agency, and emotional regulation; decrease pain perception; increase reality orientation; enhance communication; improve motor skills; foster self-expression; decrease stress; and more.

"The relationship between music, health, and overall well-being is immensely complex," Dr. Allen asserts. "Music therapists are adept at fostering collaborations between leaders in multiple disciplines to develop and implement the most innovative tools, programs, research, and practices regarding music and music-based experiences from wellness to disease management. This includes incubating and accelerating solution-focused programs and technologies focused on impact, effectiveness, sustainability, and replication within the music and health space."

Music therapy is a large and growing practice today, but it's been around, in one form or another, for centuries. In the Bible, the young David is tasked with playing the lyre (the beat-up campfire acoustic guitar of the ancient world) to calm the ever-agitated King Saul. The ancient Greeks used music to treat conditions ranging from depression to hangovers, and Plato himself called music "the medicine of the soul." But modern, professional music therapy really started taking off after World War II, as musical entertainment for convalescing wounded soldiers was discovered to be helping with treatment, too. Music had long been known to boost troops' morale, and as it began to release emotions, create connection, and reawaken distressed individuals, it became clear that music might hold all manner of curative powers beyond its entertainment value.

And so it does. In the decades since, science has unearthed mountains of evidence that targeted music interventions can improve our health all along the journey from the cradle to the grave. With more than 9,000 professionally accredited music therapists in the world today and a strong tailwind of scientific evidence at its back, music therapy is just beginning to explode into the popular consciousness and is poised to take its place beside exercise and nutrition in large-scale impact on human health.

The promise of this therapy lies in music's ability to affect nearly all the systems of the body. Your thumping playlist activates your auditory cortex, but it also affects areas involved in emotions, rewards, cognition, movement, and pain. "These brain regions have been implicated in a broad range of organ dysfunctions and nervous system disorders," points out Dr. Helen Lavretsky, a professor in residence in UCLA's Department of Psychiatry. The fact that music impacts the same regions that are responsible for disorders, says Lavretsky, provided the initial theoretical premise to start investigating the potential of music intervention to improve health and well-being.

**FUN FACT**

Two out of three people said they would shorten or skip their workout entirely if they couldn't find their headphones.

So what specific conditions and diseases do musical interventions address? Almost all of them, it turns out. In hospital neonatal units, music therapy helps premature babies relax, stabilizes their heart rate and breathing, improves their oxygen saturation and nutrition, and sends them home sooner. For patients at the end of life, music therapy promotes continued **brain plasticity**, helps Parkinson's patients move, helps dementia sufferers remember and PTSD sufferers forget, and staves off cognitive decline. And in between, it helps people of all ages beat addiction and depression, relieve chronic pain, improve mental health, and a whole lot more.

Music's power to heal starts with its ability to relax you and lower your stress. When calming music is played for children receiving an IV injection, for example, the kids feel significantly less pain, and administrators report that the IVs are easier to insert. If music did nothing but soothe patients, it would be a tool of immense value to the medical community. But that's only its warm-up act.

Leveraging the brain's extraordinary receptiveness to sound, musical interventions use rhythm as a strategic input to mechanically change the brain's outputs. Music with predictable rhythms—aka songs—can leverage the brain's natural ability to "rewire" itself (strengthening connections and building new ones) in directed and positive ways.

An example: Dr. Joanne Loewy, founding director of the Louis Armstrong Center for Music and Medicine at Mount Sinai Health System, has authored multiple studies assessing music therapy's ability to improve the health of premature infants. In one study, researchers held music therapy sessions that included both the infants and their typically anxious parents, featuring trained music therapists performing gentle music specifically entrained to each premature infant's own vital signs. The therapy achieved remarkable results, improving respiration and heart rate in the premature babies. But it also relieved anxiety in the worried parents and improved parent-child bonding. Dr. Loewy told me there's a lot more where that came from. "We are now studying music therapy with neonates [newborn babies] who have abstinence syndrome—they were born dependent on narcotics," she told me. "Music therapy assists their calm, improves sleep, and helps mothers and fathers feel connected and competent."

Even infants without developmental issues can gain significant health benefits from music.

Babies who engage in music-centered play sessions—as compared to nonmusical play sessions with toy trucks and dolls—become better at recognizing and responding to sound patterns. Nine months later, those children already have a measurable cognitive upper hand over their peers—particularly in pattern perception, which is important for speech development.

"We know that people who play music or learn to play music when they're younger are more resilient to recovery from difficult things they face when they are older," says Susan Magsamen, executive director of the Arts + Mind Lab, Center for Applied Neuroaesthetics, Johns Hopkins University School of Medicine, and codirector of the NeuroArts Blueprint with the Aspen Institute. "There's a protective value to music, whether or not you're 'good at it,' and whether you're the maker or the beholder."

Let's take a minute to explore this "protective value" of music with a brief dive into how music improves our most common active effort to improve our own health: the daily workout.

## LET'S GET PHYSICAL

"When I work out, I listen to certain music," says Steve Aoki. "It pushes me to that point of drive where I can really have the most efficient activity." We need drive, endurance, and focus when we work out because working out hurts. When you're half an hour into your morning jog or finishing your first set of sumo squats and feeling the burn, it's because your body is working overtime. The energy for your exercise comes from a compound called adenosine triphosphate, or ATP. As your muscles contract and "burn" ATP, your lungs gasp for oxygen to help produce more. Oxygen reacts with lactic acid, a by-product built up in active muscles, and creates carbon dioxide and water. Your core temperature rises; you begin sweating; dopamine and endorphins surge.

The conductor of this sweaty symphony is your heart, and the tempo of your heartbeat has direct consequences for these bodily changes. This is why personal trainers and gym rats are always squawking about your "optimal heart rate." Research suggests that certain activities are best performed—and that athletes achieve optimal results—at specific heart rates. A 5K race, for instance, is best run at around 80% of your maximum heart rate. For a moderate weightlifting session, you want your heart pounding at no more than 50–60% of your max.

This is an efficiency play: People who pay close attention to their target heart rate reap the greatest benefits from exercise. They lose more weight, gain muscle faster, and generally reach their fitness goals more quickly. And that's where music comes in, because, as you may recall from Chapter 2: Focus, you *do* have some control over your heartbeat—we know its tempo rises and falls depending on the kind of music you're listening to, if it's rhythmic and loud enough.

So the key to optimizing workout efficiency is to figure out your target heart rate for a particular exercise and listen to songs that match that tempo. Here's how to do the math. (This is for entertainment purposes only— always consult your doctor before starting or changing any exercise regimen.)

- To find your maximum heart rate, subtract your age from 220. For a 45-year-old, for example, the maximum heart rate would be $220 - 45 = 175$.

- Each exercise has its own easily Googled target percentage. In our 5K example, your target heart rate is 80 percent of your maximum heart rate. Our 45-year-old's target heart rate would therefore be $.80 \times 175 = 140$.

- Finally, to maintain that target heart rate (in this case, 140), you want to play music that also has a tempo of 140 bpm, like No Doubt's "Spider-webs," or "Womanizer" by Britney Spears (ex-amples chosen for relevance to our theoretical 45-year-old...online widgets can help you find the tempo of any songs that warm your gravy).

# Music makes exercise more efficient: Cyclists who listen to music consume 7% less oxygen than their counterparts riding in silence.

It may seem remarkable that your body literally takes its cues from the tempo of the music you're listening to, but that's exactly what happens. In one astonishing study, scientists in the UK had college students ride stationary bikes three times, listening to the same song each time. But unbeknownst to the participants, that song was played once at normal speed, once 10% faster, and once 10% slower. The result: At the slower tempo, riders exhibited a lower heart rate and rode for fewer miles; when riders listened to a slightly faster version, they literally worked harder and went farther.

Music's effects on your workout aren't limited to the rhythm of your heart. The right playlist drowns out distractions and pumps up your enthusiasm, too; people who listen to music during their workouts exercise an average of 15 minutes longer. "Pump-up" music with heavy bass causes the body to release natural painkillers, helping people push through fatigue and pain. And for endurance exercises such as distance running, music has been shown to reduce the "perceived effort" by 12%. Music even aids mechanical efficiency, somehow: Bicyclists who listen to music while riding use 7% less oxygen than those who ride in silence.

In other words, if you take the time to optimize your playlists for your workouts, you'll exercise harder, it'll hurt less, it'll feel easier, and you'll get better results.

# FINE-TUNE YOUR MUSICAL WORKOUT

*Matching song tempo to target heart rate can help optimize your workout for maximum brain-heart sync. Here are some ideas...*

| | AGE | THR* | SAMPLE SONG AT THIS BPM† |
|---|---|---|---|
| **STRETCHING** | 25 | 117 BPM | "Love You Like a Love Song," Selena Gomez and the Scene |
| | 35 | 102 BPM | "Work It," Missy Elliott |
| | 45 | 92 BPM | "No Scrubs," TLC |
| | 55+ | 87 BPM | "Blowin' in the Wind," Bob Dylan |
| **WEIGHT-LIFTING** | 25 | 129 BPM | "State of Grace," Taylor Swift |
| | 35 | 116 BPM | "Smooth," Santana |
| | 45 | 108 BPM | "Just a Girl," No Doubt |
| | 55+ | 102 BPM | "Solsbury Hill," Peter Gabriel |
| **CROSS-TRAINING** | 25 | 144 BPM | "Locked Out of Heaven," Bruno Mars |
| | 35 | 122 BPM | "One More Time," Daft Punk |
| | 45 | 120 BPM | "Orange Crush," R.E.M. |
| | 55+ | 117 BPM | "Hot Blooded," Foreigner |
| **BICYCLING** | 25 | 140 BPM | "7 rings," Ariana Grande |
| | 35 | 137 BPM | "Use Somebody," Kings of Leon |
| | 45 | 133 BPM | "Tour de France," Kraftwerk |
| | 55+ | 111 BPM | "Rapper's Delight," Sugar Hill Gang |
| **RUNNING** | 25 | 147 BPM | "Physical," Dua Lipa |
| | 35 | 142 BPM | "Mr Jones," Counting Crows |
| | 45 | 136 BPM | "Roxanne," The Police |
| | 55+ | 123 BPM | "Celebration," Kool & the Gang |

*Target heart rate    †Beats per minute

# JANE'S ADDICTION

To truly grasp the potential of music therapy, look no further than the dark world of chemical addiction. Our brain's reward system has a fundamental flaw: We don't know, by and large, when we've had enough. Eating food, exercising, giving to charity, and having crazy monkey sex all earn us a payload of feel-good neurotransmitters for a job well done. But certain drugs, including alcohol, essentially hijack the limbic system to trigger the release of up to ten times the dopamine dosage your brain would normally release. You know the feeling as being drunk or high, and you should *definitely* put those car keys in the bowl over there by the Funyuns. That's really too big a burst of pleasure, and after the substance wears off, your brain can struggle to regain its normal chemical balance, leaving you with a desire—no, a *need*—to revisit the experience, at first to get that pleasure again, and later, over time, just to get normalcy back.

Not everyone who uses hard drugs becomes an addict; only about 30% of those who try heroin, for example, get hooked. But close to 30 million U.S. adults report having suffered from a drug-use disorder during their lifetime, according to the National Institutes of Health. And the vast majority of them never receive any type of treatment.

Sadly, there's no magic playlist to cure drug addiction. But one of the biggest challenges of substance-abuse treatment is getting people to stick with the programs, and that's where music therapy can help. A 2009 study in the journal *Drug and Alcohol Review* found that, over a seven-week trial, people who participated in music therapy were more engaged in their treatment program than those who didn't. Group music exercises reportedly made participants feel less isolated and helped them connect with others; the therapy itself lowered stress, helped patients relax, and boosted their motivation.

Music therapy can also help recovering drug addicts relieve anxiety, which is no small thing. A 2013 study showed that 85% of women with substance-abuse disorders and a history of anxiety saw their symptoms decline after just one group music-therapy session. The study's authors suggested that the targeted musical intervention, including drumming and vocal improvisation exercises, "allowed for a release of physical tension leading to a decreased psychomotor agitation," while singing songs "appears to have encouraged an important emotional release, or catharsis, that paved the way for a greater sense of inner calm."

In other words, music therapy helped anxious patients to simply *let go*.

This rockin' complementary therapy isn't limited to substance disorders. Other "behavioral addictions"—such as eating, gambling, and sex addictions—can benefit from similar treatment strategies (where music therapy is combined with an overall treatment program), and the reason they work might have something to do with the way the brain organizes around habits. When scientists compared the MRI images from the brains of cocaine addicts and gambling addicts, they were surprised to find a common link: Both sets showed lukewarm activity in a reward center called the ventral striatum and in the prefrontal cortex areas involved in impulse control. Called "reward deficiency syndrome," this phenomenon implies that people prone to addiction have an underactive brain reward system and therefore have to work harder—snort more, gamble more—to get the same baseline dopamine kick.

Despite the neurological similarity to chemical dependency, behavioral disorders like these are considered cognitive conditions best treated by behavioral therapy. And that's where music comes in. Pathological gambling, for example, often co-occurs with stress, and music therapy relieves that stress, causing blood pressure and heart rates to drop, which over time might make problem gamblers feel less harried and impulsive. Music therapy can also function as a social lubricant that improves the value of other kinds of therapy. As the authors of one study on music therapy and gambling noted, "Music tended to stimulate and deepen the discussion in some sessions where discussion seemed to be trifling or troublesome." That is, music helps people break out of their shells—and that can potentially catalyze breakthroughs.

## WHEN I'M 64

Here's a fun fact I found out in researching this book: Did you know your brain shrinks as you get older, by as much as 10–15%? I'm looking forward to forgetting *that* fact in a few years.

It's our **frontal lobe**, the region responsible for logic, learning, and the coordination of thoughtful action, that sees the biggest decline. Even if we're lucky enough to avoid one of life's awful little endgames like **Alzheimer's** or Parkinson's, many of us will suffer cognitive decline, if we live long enough, as our neurons stop communicating as effectively, reaction time slows, and serial tasking becomes increasingly difficult.

It's not all downhill from here, however: Some cognitive functions actually improve with age. **Emotional intelligence**, for example, peaks in your 50s, while semantic knowledge—otherwise known as your vocabulary—peaks in your 70s. And your mental storehouse of facts and trivia, which psychologists call "crystallized knowledge" but which my kids call "dadsplaining," continues to rise far into your golden years. On a personal note, I can tell you I played Scrabble off and on for years with my great grandmother Marguerite until she went into her final decline, and I never did beat her. Imagine having a hundred words like "qi" and "xu" and "jo" at your beck and call...and yes, she could use them all in a sentence.

Nonetheless, cognitive decline is real, and can creep in slowly, insidiously, causing other problems. For example, a coping mechanism you're probably unaware of is that your auditory system normally produces a kind of standing wave that drowns out background chatter—that's how you can have a conversation in the middle of a raucous party or at a table in any New York City restaurant. But your ability to produce this wave degrades over time, making it harder and harder to separate the conversation you're

> "Listening to music interferes with your brain's perception of pain, simultaneously reducing anxiety and muscle tension while releasing endorphins."
>
> DR. JOY ALLEN *on how music therapy can ease physical pain*

# FOUR QUESTIONS FOR

# QUESTLOVE

This actor, drummer, TV star, Oscar winner, six-time Grammy winner, and cofounder of The Roots is a force of nature. The musical director of *The Tonight Show Starring Jimmy Fallon* (where The Roots is the house band), Questlove also finds time for outside pursuits... like deejaying Obama's final White House party in 2017.

**How do you think a song specifically can make us feel about history and culture?**

So here's a great example: To be in America in the last seven years, especially during the second half of the Obama era, blatant racist acts have been happening to people of color. I mean, sure, throughout history, but really amplified in a way which I guess we could say starts with Trayvon Martin, that's really when the world starts paying attention.

So I noticed that there were a bunch of artists that you could clearly see gearing up for the rebound or gearing up for a position in terms of like, "I'm gonna write 'the Anthem.'" I could see people gearing up, like "I'm gonna write the new 'Lift Every Voice and Sing.'"

And what's weird is that what winds up happening is that the person that nails it and gets it right never went in with intentions to be that, to capture the zeitgeist, or the lightning in the bottle.

So in the case of Kendrick Lamar, "To Pimp a Butterfly," he winds up really creating what I believe is the Black Lives Matter anthem, which is a song called "Alright."

It's not like they were calculating in there, "We're gonna have an anthem thing." But I think at the end of the day, the moment chooses the anthem and not the anthem choosing the moment. And I think that's as good as it can come. And I've been guilty of this, the amount of times where a filmmaker has asked us to create a song that is gonna go up at the end credits. And then the first thing you start thinking about is your acceptance speech at the awards.

**Do you have a different emotional connection to a song writing, performing, or spinning?**

I think I actually get more joy deejaying music. I think that playing music, there's a magic there that's just as strong as the person who wrote the song.

I've never cried creating a song. I mean, I've had goosebumps creating an album. But I will say that there have been times—I think four or five times—where I've spun a record and I saw its response to an audience and that made me cry as if I wrote the song.

**How do you think your creativity impacts your music differently than other projects you produce?**

In the beginning, my goal was to be the best drummer. And then one day I outgrew that and I was like, I wanna be the best band leader. Like I want my band to be awesome too. And then once we conjured up that magic, then I wanted to know if that magic can spread to other areas without my band. So then it's like, I wanna be a great producer, but then it was much more than that. I was an organizer...I always do this 10-year pivot thing.

So I started studying creativity and now I have to put that to use in a way that I'm not used to before, now that I'm 50. I've spent the first 50 years of my life playing in the shallow end of the pool of things that I know. I know records, I know food. I like to think I'm funny, I know comedy. But I didn't know directing.

**How did you build your creative chops in other creative industries?**

The one thing that my producer told me when I started my movie, he says, "Well, this is what I want you to do. You're not directing a movie. This is a deejay gig now, what are you gonna do?"

So he gives me 40 hours' worth of concert footage. And he is like, find 30 goosebump moments that really just shake you to your core of amazingness. And those are your records. And so I spent five months on loop digging, and, you know, I've realized that this whole entire, even the way that we edit it, the way that we cut it, that's literally me as a deejay.

What I'm saying is I think I'm in the business of creativity. I guess my goal in life is to encourage other creatives that aren't exactly married. I'm like the polyamorous version of creativity.

# "There may be teardrops to shed, So while there's moonlight and music and love and romance, Let's face the music and dance."

**TONY BENNETT** "Let's Face the Music and Dance"

having from the background noise. "As it becomes more difficult to understand speech in noisy environments with aging, people can stop socializing, as it is too frustrating," Laurel Trainor told me. "Social isolation is a serious problem in aging populations." Heather Read echoes this sentiment, adding: "This inability to suppress irrelevant sensory information likely accounts for some of the impaired social cognition, attention, and memory evident in dementia."

For otherwise functioning older people marginalized in this way, musical activities like singing in a choir can help draw them back into that protective social fold. But sometimes, cognitive decline can be so pronounced that we start to lose the person themselves. This is the world of Alzheimer's disease, a terrible condition where runaway brain plasticity produces impenetrable tangles, leading to an increasingly isolated and often scared individual who no longer recognizes the loved ones trying desperately to reach them.

And here's where music's powerful connection to memory can work some true magic.

## YOU MUST REMEMBER THIS

I was fortunate enough to see Tony Bennett and Lady Gaga perform a few years ago; they were spectacular. The pandemic brought a cruel end to Bennett's miraculous late-in-life touring, and when it was announced, in early 2021, that the 94-year-old legend was suffering from Alzheimer's disease, it was surely the end of a remarkable career.

Only…it wasn't. Not quite yet.

When Bennett sat for a *60 Minutes* interview in late 2021, just months after the announcement, he was already in noticeable decline. He sat mostly motionless, eyes distant, as his wife and caretaker, Susan Crow, did the talking. When prompted with a simple-enough question, Bennett could deliver the briefest of responses, but it seemed like the end of the road.

Then, from the next room, his longtime musical collaborator started playing the piano. It was like a switch flipped in Bennett's head; he strode over to the side of the piano and started belting out his classics, one after the other. The interviewer, Anderson Cooper, says Bennett sang an hour-long set entirely from memory. Tony would go on to play two more sold-out shows at Radio City Music Hall, performing brilliantly, remembering his costar, Lady Gaga, by name, and interacting with the crowd. He seemed happy and relaxed, 100% in his element. Later, he literally didn't remember having performed at all.

Tony's neurologist, Dr. Gayatri Devi, explained: "It's his musical memory and his ability to be a performer. Those are an innate and hardwired part of his brain. So even though he doesn't know what the day might be, or where his apartment is, he still can sing the whole repertoire of the American Songbook and move people."

For reasons that scientists are still exploring, Alzheimer's often causes a person's cognitive functions to deteriorate but spares (at least at first) their ability to recognize music. The leading theory suggests this is because music engages so many parts of the brain that musical memories are encoded multiple times, and thus strengthened with repetition and association—like Bennett's countless public appearances over the decades. Even as copies of music stored in the brain's **episodic memory** (past events) deteriorate, the same music locked

## How does dancing help with Alzheimer's?

*Music engages the motor system. It's the reason, when our playlist is bumping, we just can't help but twist, shimmy, or shake. And this complex brain pathway may also help Alzheimer's patients. A review of 349 Alzheimer's patients found that dance therapy was effective in maintaining a person's quality of life as the disease progressed. Astonishingly, it might even be a factor in preventing Alzheimer's altogether. A 21-year study published in the* New England Journal of Medicine *showed that physical activities like swimming and biking, while good for health generally, essentially did nothing to prevent Alzheimer's. But dancing did: In fact, frequent dancers reduced their Alzheimer's risk by a whopping 76%!*

inside a person's **procedural memory** (how to do things) might remain relatively strong, and so on.

And because music enjoys this unique protected status in our minds, it can be a powerful therapeutic tool. "Personalized music programs can activate the brain, especially for patients who are losing contact with their environment," said Alzheimer's expert Norman Foster. In 2019, Foster and other researchers used functional MRIs to observe the brain activity of dementia patients; when these patients listened to their once familiar tunes, their brains lit up like the proverbial Christmas tree. "This is objective

# Playing a 40 Hz buzzing sound for mice in an Alzheimer's study helped clear plaques in the brain and improved the mice's memory.

evidence from brain imaging that shows personally meaningful music is an alternative route for communicating with patients who have Alzheimer's disease," Foster said.

Dementia causes dramatic changes to the brain; for sufferers, the world they once knew can start to look and feel foreign. It's a massively disorienting experience that can cause confusion, depression, and anxiety. Music can be a familiar anchoring point that reawakens them to the value of the world and keeps them engaged with it, making dementia sufferers prime candidates for music therapy. "Music intervention reduces associated depression and improves cognitive function and quality of life in people with dementia," says Heather Read. Song by song, music therapy is breaking through barriers and improving elderly patients' lives right now.

To boost the spirits of an Alzheimer's patient, for example, music therapists might use something called the "**iso principle**." First, they take a song that matches a person's current mood, then they build a playlist that progresses toward a more desired state of feeling. (You might remember this two-step pattern from our advice on getting over a breakup in Chapter 3: Love.) "If someone is feeling sad or has low energy and we want them to feel happier, you start with music that validates their feelings," Sarah Folsom, a music therapist, told *TMC Innovation,* the Texas Medical Center blog. "It's their own preferred music—so whatever you listen to when you're sad—and then you slowly have the playlist move towards songs that are more uplifting to you."

Music's connection to memory can also be leveraged to help dementia sufferers remember things they'd otherwise forget, in much the same way we teach children their ABCs by encoding them in song. "Several studies have shown that music can be effectively used as a mnemonic tool...in particular [for] Alzheimer's patients, who have been shown to learn and retain information better when they are sung rather than spoken," says Dr. Laura Ferreri, associate professor of cognitive psychology at Université Lumière Lyon 2.

As we get older, as our brains shrink and our memories fade, our soundtrack remains. Even in the tangle of Alzheimer's, the music we've absorbed and loved stays with us until our dying day.

## CONCLUSION

The gathering evidence that music can play a key role in healthcare as a universal, easy, low-cost, complementary therapy is transforming patient outcomes—and driving big changes.

I spoke to Susan Magsamen, codirector of the NeuroArts Blueprint, an initiative with the Aspen Institute, about this global effort working with researchers, artists, policy makers, and funders around the world to advance the science of the arts and elevate the field using music and sound as viable health and wellness interventions. "There is compelling research around the arts showing that people who have more experiences like concerts live longer lives, after accounting for lots of variables, and they're sick less," says Magsamen. "In the future, you're going to see arts practice become part of how we manage our health."

Others are exploring how technology can be used to help amplify music's healing powers. Adam Gazzaley, M.D., is the cofounder of Akili Interactive Labs, creators of the first videogame approved by the FDA as a treatment for a medical condition (ADHD), and a sponsor of various other initiatives exploring ways to improve human performance through technology. "What I do is look at real-world experiences where there's evidence they can improve cognition," he told me, "and then figure out how they can be delivered through technology to allow them to be more accessible, more scalable, and turn it into medicine."

From the beginnings of human civilization, music has been instrumental (sorry) in helping us heal. It lowers our stress, brings us pleasure, and connects us socially—all factors in setting a healthy baseline. The rhythms of our songs can fine-tune our heartbeat and brainwaves to achieve the outcomes we're looking for, from great workouts to reduced anxiety and depression. And their deep connections to memory help anchor us to our past, keeping us solidly connected to our history and identity, even when our cognition starts to fail us.

The bottom line: Music is an extraordinarily powerful and promising tool for healthcare, and one we're only just beginning to learn to leverage. ∎

# TAKEAWAYS

*The world is a stressful place, and it can be hard to stay healthy. But music can help: It can lift your spirits, keep you motivated, improve your workouts, and help you climb out of the depths of depression or addiction. Even if you've been taking your body for granted in the usual 21st-century ways—not enough working out and hydration, too much drinking, smoking, and donut-eating —music can help you flip the script and replace your bad habits with good ones. And music therapy is exploding with specific, science-backed applications for a vast array of disease states, from neonatal health to end-of-life dementia. Here are some places you can leverage music to uplevel your health.*

**TO MAKE WORKOUTS MORE EFFICIENT** Any music can provide the right mix of inspiration and distraction to improve your workouts, when played loud enough to eclipse background noise (or through headphones). But if you play music that matches your target heart rate for a specific workout activity, you'll work out harder, build muscle mass quicker, feel less pain and fatigue, and reach your fitness goals more quickly. And maybe have more fun.

**TO COMMUNICATE WITH DEMENTIA SUFFERERS** Try finding familiar-to-them music as a means to forge a connection that still makes sense to them, hacking into their procedural memory to keep them engaged.

**TO DEVELOP HEALTHIER HABITS** Try group music therapy in coordination with a relevant professional program (for example, a substance-abuse program). It'll make you more likely to stick with the program long enough to achieve results.

CONNECT

# SWEET CAROLINE NEVER SEEMED SO

**NEIL DIAMOND**
"Sweet Caroline"

**\*SO GOOD!**
*SO GOOD!*
*SO GOOD!*

# GOOD TIMES

# GOOD

**IT WAS ITALY**—sweet, sun-drenched, eternal, kiss-you-on-both-cheeks Italy—that bore the brunt of the first wave of the COVID-19 pandemic. By March of 2020 the virus had been percolating in the country's north for weeks; as it spread south and body counts climbed, the government reluctantly instituted an unprecedented countrywide lockdown. Flights were suspended, sporting events were canceled, schools and churches were shuttered. And for the first time in living memory, the people of Italy were prohibited from leaving their own homes.

Sixty million gregarious Italians fretted and fumed inside their houses and apartments, as the country's fabled cities became picturesque ghost towns where endless sirens echoed down empty cobblestone streets. For this tourist-dependent country that prizes family and community above all else, the pandemic was the worst of all possible worlds.

But then, something incredible happened...People started to sing.

In city after city, night after night, Italians stepped out onto their individual balconies to join their neighbors in a chorus of song. Sometimes it was Italy's national anthem. Other times, an opera diva would belt an aria from her window, or the neighborhood *nonnas* would grab their pots and pans and bang out a beat on the balcony. The music spread faster than the virus; soon people as far away as Spain and Sweden were joining the chorus. Watching YouTube videos of Italy's musical defiance became a worldwide obsession.

These impromptu performances did a lot more than just break up the monotony of quarantine. They provided moral support for shellshocked frontline workers. They reinvigorated Italy's national pride. And they helped confined citizens reestablish their local communities, as neighbors—even apartment building tenants who had never met in person—shared the stage to provide a beacon of hope in one of their country's darkest hours.

Behold, the power of music.

Humans have a need to connect that's every bit as real as thirst or hunger. We are defined, in no small measure, by our relationships: our friends and families, our classmates and coworkers, our lovers and neighbors, our teachers

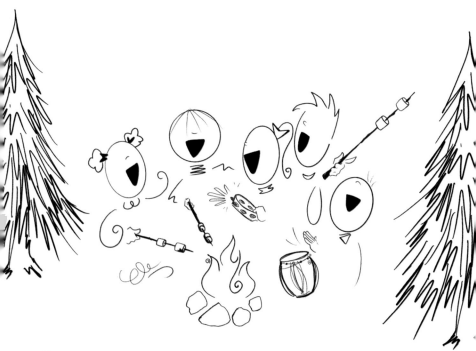

> "One of the theories is that music started not only as the way of uniting people, but as a way of uniting people to kill each other. So it was a way of uniting people in battle to totally identify with your tribe, with your group."

TOD MACHOVER *on music and your posse*

and therapists. Our global civilization has become a complex web of interdependence, where food from around the world shows up magically in our stores and garbage disappears magically from our curbs. This interdependence is hardwired into us now; we generally assume that the people around us can be relied upon to provide continual resources and ongoing support, a phenomenon psychologists call **Social Baseline Theory**.

And one of the best ways to forge and strengthen these critical connections, as humans have literally always known, is through music.

## THE SONG REMAINS THE SAME

Music has been with us since sometime before the dawn of civilization. When we search ancient cities, we always find evidence of music, including some of the earliest human relics ever found: alligator-skin drums and flutes carved from cave bear bones that date back 50,000 years, for example. "There is no known culture now or anytime in the past that lacks [music]," says Daniel Levitin in *The World in Six Songs*.

FOUR QUESTIONS FOR

# YOUSSOU N'DOUR

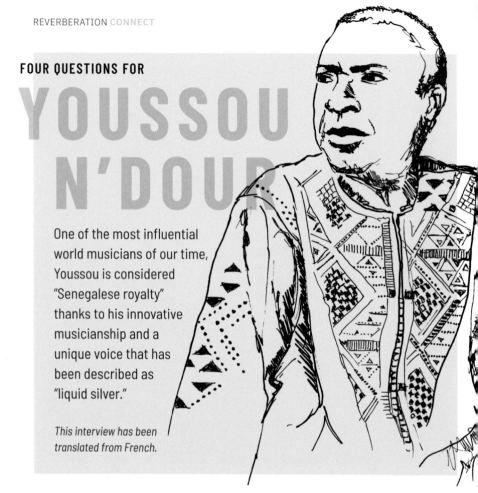

One of the most influential world musicians of our time, Youssou is considered "Senegalese royalty" thanks to his innovative musicianship and a unique voice that has been described as "liquid silver."

*This interview has been translated from French.*

**Why is music important?**

I think that first, music for me is family. I was born from a griot family on my mom's side, and music always puts me back into my history. When I started playing music, I was very young, I was 13, and would watch my grandmother sing. That close relationship with my grandmother gave me the opportunity to take music to the next step.

**Do you think about how your music will affect your audience while you are writing it?**

Yes, I think that melody-wise, I sometimes feel the relationship between a singer and his audience is about melody and the sound of the voice.

I hope that at the next phase of writing, I come up with a melody that sounds interesting enough on its own. The words will then prepare the audience and take them toward that melody, which may be a little bit strange, but will connect well with the lyrics.

When I'm writing, first it's a melody—it's something that happens to me. Sometimes the inspiration hits me in the shower, but most times in the evening when I am alone.

Usually I try to capture it in the moment, and if I can't, I do it later on—which means it's a good idea. A good idea doesn't need to be recorded in the moment, because if it's a good idea it stays in your head, it will come back to you.

The most difficult part is to decide the theme of what I want to say. It's much harder to find the words of what I want to say.

**Do you think music can transcend language?**

For me, first of all, music is a first language. It is a language that everyone understands without being translated. If you don't understand music as a universal language, then you cannot understand somebody from another culture. Before the words come in, it's the voice and the emotion that goes with it, and that can touch someone who doesn't understand your language.

Music can amplify emotions and bring people together...like when I listen to a beautiful melody like something from Bob Marley, or "Biko" by Peter Gabriel. At first I don't understand the words, but like a lot of people, I replace the words with something I know, because I am touched by the melody of the song and the emotion created by the song, so the words are secondary.

**What role does music play in your daily life?**

Once I became passionate, it was like a driving force. When someone speaks, I always feel inspired, as if it were a song. When I hear people speak or hear the Muslim prayers in the morning, it's all translated to music. So my entire environment and everything around me is music.

When you are that passionate, then all you hear is music. When I hear music on the radio or other places, I take stock in everything that influences me during the day, and then in the evening, I try to find the response through music to calm me down.

For me, everything I listen to is music. Sometimes when people are talking to me, I travel in my mind because it sounds like music—then I will come back and continue the conversation. I don't listen to music like normal people do, because everything I hear is music. If there is no music around, I don't feel free.

But music wasn't just along for the ride: It seems to have helped create civilization itself.

Making music is a fundamentally collaborative exercise, in which some people write or compose the music, others perform, and still others listen or dance. (With some overlap, of course.) "What you're really doing is interacting with the rest of the band," says Pink Floyd's drummer, Nick Mason. "The interesting thing is it doesn't matter if it's 200 people in a pub or 200,000...what works best is to almost play for each other." Live music—the only kind, until relatively recently—is a shared experience that plucks people out of their daily grind and immerses them in a temporary shared space that lives outside normal reality. Other kinds of art, like plays, can do that, too. But music goes further, producing specific chemical changes in the brain that actively promote social interaction and connectivity.

**FUN FACT**

# Six in 10 Americans say a day without listening to music is worse than a day without human interaction.

Music has been found to strengthen social bonds in four specific ways, according to UC Berkeley's Greater Good Science Center:

- It increases contact, coordination, and cooperation with others.

- It provides an oxytocin boost to both performers and listeners, chemically producing "prosocial" feelings of trust and positivity.

- It activates empathy as we try to understand the intent of the songwriters and performers, and what's being communicated.

- It increases cultural cohesion, by communicating a sense of belonging and obligation toward the group.

Taken together, music's gifts dramatically enhance our ability to build relationships with other people.

The release of oxytocin is interesting. An ancient neuropeptide found in all mammals, oxytocin is responsible for that special high you feel when you hold and snuggle a puppy—or your newborn. It plays a critical role in the birthing process and in breastfeeding, it's released after sex (helping convince you both to stay the night), and it gets parents to generally pay closer attention to their offspring. Men who receive a nasal spray of oxytocin are more likely to play with infants, and pregnant women with elevated oxytocin levels during their first trimester are more likely to feel closer to their babies after they're born.

Providing chemical rewards for good parenting helps ensure that infants get the loving support they need. Even something as innocuous as singing to your children can deliver mission-critical connectivity. "Infants are not very good at regulating their state, and caregivers use singing to help them to do this—calming them when they are upset and tired, and attracting their attention when they are ready to play and learn," says Laurel Trainor, professor at McMaster University. "This early social interaction is really important for infants to feel secure, and the attachment to a caregiver is critical for physical, social-emotional, and cognitive development."

Music's ability to produce that oxytocin reward can help explain its powerful and ubiquitous role across human history. As Alan Harvey, the author of a recent study of oxytocin's role in human musicality, put it, "Music encourages affiliative interactions in infancy and adulthood...Music and its evolutionary partner dance also promote synchrony and social interaction, contribute to cultural identity, and encourage the formation of cooperative networks."

Music promotes brain changes that make individuals more receptive to connecting—so far so good. But in a group setting, music takes the connective magic to another level, providing something called **social flow**.

The flow state—as discussed in more detail in Chapter 2: Focus—is what happens when you're "in the groove" and so deeply immersed in an activity that all distractions peel away. But while flow state is an intensely personal phenomenon, social flow is what happens when multiple people experience flow simultaneously.

Social flow is common among musicians: Researchers tested jazz singers while performing previously composed music and while improvising, and found that both activities reduced stress and significantly increased social flow, that sensation of being completely immersed in an activity with others. "Jazz musicians frequently report feelings associated with flow experiences such as intense oneness with their musical product and the merging of individual musicians to form a single cohesive entity when performing," wrote the research team, led by Jason R. Keeler.

Improvisation is an example of social flow, where trained musicians can go off the playlist and create new unrehearsed music live, taking individual musician-specific journeys that somehow add up to one coherent listening experience. "Improvisation to me is like language," says legendary jazz virtuoso Branford Marsalis. "You do a rigorous amount of training, and then when the training's done . . . it's like a conversation you have with your friends. The myth of improvisation is the mistaken assumption that you can play anything, but what you

are playing has to be relative to the people you're playing with, which means that you have to have a tremendous sound palette if you're going to play with an orchestra and then suddenly play with Sting or play with Dizzy Gillespie."

Social flow can happen to listeners, too. When the entire audience is captivated and lost in the music at a concert or a Broadway musical, that's social flow. The applause at the end of a spellbinding song can be viewed as the grateful audience signaling that they've "snapped out of it" as a group and are back in reality, thanking the musicians for having taken them away.

To understand how social flow like that generated by music can connect people, let's take a closer look at music's surprising relationship to language.

## BIRD IS THE WORD

At the end of the 18th century the idea that the first language used by the human race was singing, and that men may well have learned it from birds, was widely spread among philosophers. Similarities between the way humans speak and birds sing have been apparent for centuries; Darwin himself declared birdsong "the nearest analogy to language." Modern research has uncovered that the brains of singing humans and birds share some

remarkable similarities, leading some to hypothesize that it's really no coincidence at all.

Some linguists argue that human language may have begun about 100,000 years ago with the imitation of the vocal cries of birds. According to this theory, humans gradually improved upon these cries and turned language into something vastly more complex, but our language retains some expressive qualities—namely, its rhythmic and singsongy character, called "prosody"—borrowed from birds. Furthermore, as language developed, that other great human communication system—music—may have developed in tandem.

When a male wren sings to its female partner, the brain of the male bird will synchronize with the female as she starts to sing, allowing both birds to essentially sing as one—the perfect duet. Researchers at the Max Planck Institute for Ornithology, who have recorded almost 650 bird pairs, explain the phenomenon thusly: "The call of the partner bird triggers a change in neuronal activity in the bird that began singing. This, in turn, affects its own singing. The result is a precise synchronization of the brain activity of both birds." It just might be the closest thing science has found to telepathy.

It happens with humans, too. When a chorus performs, the heartbeats of individual singers slowly synchronize, because they can only breathe in the spaces between lyrics, and as their breathing syncs up, this drives micro-changes in pulse rate. Choral singing produces a host of positive effects for individual singers, including a "singers' high" of dopamine

pleasure and serotonin contentment. It reduces stress-producing cortisol, lowers blood pressure, and improves immune-system function. And it makes the singers sharing this experience feel close to one another while increasing pleasure and reducing stress for listeners. Everybody wins.

to feelings of solidarity with comrades," says Laurel Trainor.

Singing together evolved right alongside language as a distinctly human communication channel, and it's as powerful today as it ever was.

You don't have to join the Mormon Tabernacle Choir to experience these effects. After all, in our lives when we do sing it's usually together: campfire singalongs, drinking songs, hymns in church, and so on. When we indulge in this ancient habit we feel connected and like-minded, which improves our commitment to our group—picture singing your alma mater's fight song with classmates, or the crowd singing the national anthem before a sporting event. "Whether at a party, a wedding, a political rally, or in the military, sharing music contributes

Interestingly, it may be that our big brains evolved not to handle sudoku and tax forms but specifically to help us manage ever-larger social networks. It's been estimated that 60% of human conversation is spent talking about personal relationships and interpersonal gossip. And the more people in your network, the more time you must spend in this way if you're going to keep all your relationships stable. This "social grooming" requirement puts a functional limit on the number of people who can be in your tribe or clan or Hogwarts house. But there's a loophole:

*Why are children better at learning music than adults?*

*"Because they're not as afraid of mistakes," says Dr. Joy Allen. Children aren't held back by the fear of looking foolish; they also have more free time, fewer distractions, and fewer preconceived notions. By adulthood, your prefrontal cortex is much more fully wired, and you may be "set in your ways" and resistant to the full immersion required for top-notch musical talent to flourish. The child next to you in the same course may be better able to present an open mind and get into a flow experience, and may retain information better.*

Bigger brains can help you handle more social-group complexity.

Some fascinating research done in 1992 showed that primates with larger **neocortex**-to-total-brain ratios (let's call this ratio N2B) were able to handle larger social groups, along a predictable range. Relative hermits like Tamarin monkeys have a small N2B ratio, and can support social groups of only about five individuals. Human brains, at the other extreme, are around 75% neocortex, and according to this theory should each be able to handle about 150 people in a stable "tribe." This has been dubbed *Dunbar's number,* and while it is only a broad average, it's been found to roughly correlate with a range of human social phenomena, from the size of ancient Roman troops to the number of individuals in people's mobile-phone networks.

Putting it all together, higher N2B ratios allowed higher numbers of stable social connections, conveying a survival advantage—there's strength in numbers. But to manage these dozens and dozens of relationships, including fostering a shared understanding of other humans' motives, needs, and reliability, humans needed new tools. One such tool is language—a critical development for helping to make interpersonal understanding vastly more efficient. But another is music, which united people in shared trust-building experiences that had no small role in promoting human connectivity and success, and gradually stabilized ever-larger groups (picture an Ariana Grande concert of 50,000 fans singing in unison) to drive civilization onward.

"Music could serve as a real glue that helps us in connecting with others," says Laura Ferreri. "When we are infants, to communicate in a pre-linguistic manner, but also when we're adults, as a real social bonding tool."

## WHY CAN'T WE BE FRIENDS?

Performing music with others helps improve your ability to understand and predict other people's mental states—a social skill called *theory of mind*. This is the basis of empathy, which is critical for making meaningful connections to other humans, and music serves as a powerful catalyst.

In fact, music is sometimes all you need to foster bonding. One research paper published in *Frontiers in Psychology* found that simply "coordinating your actions with a complete stranger through participation in a musical game" is all it takes to achieve "close-friend" level empathy. But why? Why should two strangers brought together randomly to play a musical game get emotionally

"Sittin' here restin' my bones

And this loneliness won't leave me alone."

OTIS REDDING "(Sittin' On) the Dock of the Bay"

# Infants remained calm twice as long when listening to a song—even an unfamiliar one—compared to listening to speech.

closer to each other? Because coordinating physical action requires subtle real-time reasoning about what the other person is doing and is likely to do next, decreasing what the study's authors called "the perceived psychological distance between individuals."

Humans are very, very good at coordinating physical behavior—picture a triple play in baseball, or children playing jump rope. Daniel Levitin's research shows that, remarkably, two people can synchronize their finger tapping on a table much faster than either one can to a metronome. It seems counterintuitive—the metronome is infallible. But with two humans, each is affecting the other's rhythm; they find a middle ground more quickly than one human can match the metronome.

Playing music or singing together forces us to look outside ourselves. Group singing is better when singers can predict how their fellow singers are likely to

behave, and this process of focusing on others makes us immediately more empathetic and more supportive of the group we're part of.

Why do we all stop what we're doing, put our hands on our hearts, and sing the national anthem before sporting events, for example? Because as a culture we've decided to leverage singing as a mechanism for connectivity, pulling the crowd together in a quick shared ritual to remember what binds us before we enter this mock battle. As further testament to the power of music to effect change, a study of UEFA Euro 2016 soccer tournament participants found that teams that sang their national anthem more passionately conceded fewer goals and were more likely to win their games in the knockout stage.

Music for the win!

# STEP TO THE BEAT

If you find it hard not to tap your feet along with a good song, you're not alone; musical behavior is closely connected to motor behavior. In fact, a theory known as **Shared Affective Motion Experience (SAME)** proposes that auditory music signals are understood by our pattern-hungry minds not as random abstractions but as "a series of intentional, expressive motor acts, recruiting similar **neural networks** in both agent and listener."

According to this theory, it's the synchronization of those neural networks with other people's (at a concert, say) that produces social bonding—the sense that you and your fellow audience members are friends and can trust one another, and that you and the band are one as well, creating this complex experience together. It's a powerful illusion.

There is mounting evidence that music may have evolved not just as a general communication tool but more specifically to facilitate long-term group living, and there's

## FUN FACT

Four in five dads say listening to music with their kids made them happier. The wheels on the bus went round and round too many times for the other one.

### *Why is music such a big deal for teenagers?*

*When you're heading into the horrible petri dish that is high school, on the edge between childhood and adulthood and trying to sort out who you are and who you want to be, music allows you to painlessly explore adult themes, and musical preference provides a low-risk shortcut to identity. "Fourteen is a sort of magic age for the development of musical tastes," Daniel Levitin told* The New York Times. *"We're just reaching a point in our cognitive development when we're developing our own tastes. And musical tastes become a badge of identity."*

# Oh, What a Lonely Boy

What about the flip side of connection—loneliness? Loneliness is a personal health disaster whose physical toll has been estimated to be equivalent to smoking 15 cigarettes a day. In 1972, French adventurer and scientist Michel Siffre explored this concept by shutting himself in a Texas

cave for six months. After a couple of months, he later reported, he could hardly string thoughts together and had lost the ability to track the passage of time; by the five-month mark, he was so desperate for companionship that he tried, unsuccessfully, to befriend a mouse.

Isolation may literally make you smaller, shrinking your prefrontal cortex (which helps you think), amygdalae (which help you feel), and hippocampus (making it harder to regulate the stress hormone cortisol). Incarcerated people in long-term solitary confinement routinely report confusion, changes in personality, and episodes of anxiety and depression. And the effects accumulate: Research has shown that babies who don't get "contact comfort" (i.e., loving touch) in their first six months of life are more likely to grow up with social and behavioral problems. Lonely people suffer higher rates of infection, cognitive decline, inflammation, and cardiovascular disease, and they are 50% more likely to die prematurely.

no shortage of evidence that it serves the same purpose today. It's been shown, for example, that doing group musical activities with children makes them more cooperative generally. For older humans, going to just one concert increases your feeling of connectedness by 25%, in addition to improving self-esteem and mental stimulation and providing other positive benefits. Going to see a show every two weeks, one study's author proposed, could add 10 years to your life.

Dancing takes it a step further. "Synchronous movement is a powerful agent for social bonding," says Laurel Trainor. "Studies in adults show that when people move in synchrony with each other (whether dancing or other movements), afterward they like each other more, trust each other more, and, if given a game to play in which they can choose to cooperate or compete, they are more likely to cooperate." Coordinating movement through music has been shown to increase both our sense of community and prosocial behavior.

## WE ARE FAMILY

In my own extended family, music is the glue that holds multiple generations together. Our family events have a soundtrack—an eclectic mix of music's greatest hits from the past 50 years: some classic '60s soul for the grandparents, a healthy helping of '70s R&B and

FOUR QUESTIONS FOR

# LAURIE ANDERSON

Laurie Anderson is a renowned avant-garde performance artist whose genius spans the art and music worlds. This electronic music pioneer invents new musical instruments and has long been known for pushing artistic boundaries.

**Do you think that music can touch you emotionally more than other artistic forms of expression?**

I've never been somebody who is interested at all in comparing; is it better with your eyes or your ears, or if you had a choice of losing your eyesight or your hearing—which would it be? I don't put things in hierarchies, ever. I think other senses that have no art forms connected to them—haptics, for example—can be more powerful than music, depending on the music. I wouldn't say music is a giant category. I have to say, I love all music with one exception, which is musicals. I don't necessarily see the value of putting different artistic forms in a kind of hierarchy. Like, what's better? What's best? I know that we're a culture that likes to find the "10 Best" or something. I'm not about that particularly.

Another sense that has no art form connected to it is smell, which also can be incredibly emotional.

People have a different reference point to the note C, and if it happens in an opera, it might not hit them the same way it happens in a pop song. So, it's very complicated.

126

**How does music hit you?**

How does music hit me right now? I'm working on an orchestra piece. And so I'm trying to work with the resonance of certain words against certain notes. It's a pilot's log of Amelia Earhart's flight around the world. So I try to find what hits me about a certain word, like "circumference," and how you can put that into a musical world without sounding ridiculous—how to keep it musical, how to keep it resonating and not just sitting there as a dead thing with a definition, but how it lives in a piece of music. So, it depends on what kind of music you're talking about.

When music hits you, I think half of it is your own willingness to be hit. And your own point of resonance, where you would like to be. You can play a really lonely song to someone, and they're going, "Boy, that's a peppy thing." Depending on your mood, one person can feel, "That's an upbeat thing," and somebody else will be on the floor sobbing. As listeners, we're collaborators.

**Are there particular times of the day when music must be on?**

Right now, I'm listening to only Philip Glass. Why? Because I'm deejaying an ice-skating party, and it's his 85th birthday party. And I can tell you that every single thing that Philip Glass ever wrote can be ice-skated to. [laughs] Every single one. It's twirling, it's turning, it's gliding.

So how do people relate to that music? I'm pushing them into this way of expressing themselves to this music, in terms of the sequence that I'm doing, and some of it is you just have to be like a hockey player and do the pendulum swing. Other times you'd have to do a twirl. So music gets into your body in a different way than other art forms. For example, I very rarely see people dancing in front of their favorite painting.

**Can you describe the difference between telling a story through instrumentation and through lyrics?**

I think that there's nothing that isn't a story. So, your face right now is a story for me, your expressions are—the way you're both wearing pink—that's a story. I have different requirements for what a story is.

I feel like I'm more of a landscape painter, to see this and that. When I'm called upon to make a kind of cause and effect or narrative stream, I tend to go back on sense impressions rather than action. I'm a believer in karma; I do believe that actions have reactions. I would say, right now my collaborator is an AI super-computer, and we're writing a lot of things together.

127

mellow classics that my wife and I trained our kids on, rockers and ballads brought back to life by *Guitar Hero,* a smattering of modern crooners like Adele and Diana Krall, and more. That constant background memory stimulation brings out the stories, bathing everyone in reflected nostalgia. It's the background music of our family's life.

We don't get any argument from the kids—our songs are also theirs, now, although their songs are not our songs. And I know now that the repetition, Thanksgiving after Thanksgiving, birthday after birthday, graduation after graduation, strengthens these memory pathways to the point where those classic songs now mean family in the same way that "Jingle Bells" means Christmas. I fully expect that our kids will associate these songs with my wife and me long after we're gone—and that is all the immortality I need.

Music's amazing ability to foster social bonds like the ones in our family may very well account for why this art form is so pervasive throughout history. The psychologist Bronwyn Tarr goes so far as to suggest that music's ability to bond strangers "may have played an important role in the evolution of human sociality."

In other words, we're not musical because we're social—we're social because we're musical. Music flourished thousands of years ago because it taught us to work together, because it provided chemical reinforcement for empathy and partnership. It strengthened social connections between strangers right when our survival depended on creating wider social networks. And it's been a central part of our shared culture ever since.

Please don't stop the please don't stop the music.

# TAKEAWAYS

*"Music is a fundamental part of our evolution; we probably sang before we spoke in syntactically guided sentences," writes Georgetown professor Dr. Jay Schulkin in an article in* Frontiers in Neuroscience. *We're born to connect with others, and our social connections provide a range of benefits, including lowering anxiety and reducing depression, helping us regulate our emotions, raising self-esteem and encouraging empathy, and even improving our immune systems. Whether it's listening to shared music at a concert, singing together in a choir or a dive bar, or performing as a group, music's natural power to connect people and facilitate communication offers no end of opportunities to unify groups.*

**TO UNITE YOUR TRIBE**
Listen to live music together. The shared rhythmic experience builds empathy and lays a strong foundation for long-term trust and cooperation.

**TO BECOME A MORE EMPA-THETIC PERSON** Consider choral singing, which synchronizes your activities with the brains of others and trains you to focus on the actions and emotions of others, and providing chemical rewards and health benefits.

**TO GET CLOSER TO STRANGERS** Find a way to perform music together. The company group-bonding experience will have a greater chance of success if it involves music: Performing a karaoke duet or writing a song together activates rhythmic cooperation and puts the parties literally in sync.

REVERBERATION

ESCAPE

# SOME PEOPLE THE PHYSICAL DEFINE

# AND I'VE BEEN THERE BEFORE

**ALICIA KEYS**
"If I Ain't
Got You"

# THINK THAT THINGS WHAT'S WITHIN

**ARE YOU FAMILIAR WITH A HORRIBLY CHEESY 1976 SONG** called "Afternoon Delight," by the Starland Vocal Band? When that song came out, I was 10 years old and on vacation with my extended family in Topock, Arizona, a speck of a town on the California border. It was in the 70s, and responsible parenting was still in its infancy: My cousins and I were left to ourselves essentially all day long while the adults drank tequila sunrises or put car keys in a bowl or whatever the hell else adults were doing. And together our unsupervised little cousin-gang did all kinds of unsafe things, like squishing bugs with our bare feet and jumping into the Colorado River from a rope swing. In particular, we spent a lot of time at the local town's lone public building: a combination restaurant, post office, bar, and pool hall.

I have no idea why we took a particular shine to that song—we certainly had zero idea what it was actually *about*. But we would hang out for hours and hours, knocking balls into pockets and playing just that one song over and over and over on the jukebox, like the blissful young idiots we were. The repetition burned it into my brain, and all my life, decade after decade, that song has never failed to take me right back there. If you're 30 you may know this song from the memorable four-part harmonizing bit in Ron Burgundy's office in *Anchorman*; if you're 40 you might remember the campy, inappropriate uncle-and-niece singalong from *Arrested Development*. For me, whenever I hear that song, I'm 10 years old, shooting pool in a bar with my cousins, daring and foolish and carefree.

"You know how they say you can smell something and it brings you right back?" says Mover. "Same thing for me: I can hear particular things and it takes me right back to my childhood. I don't know if I can tell you why I listen to music all the time, but I can tell you that it's an incredibly comfortable place to be, to always have music going, regardless of what kind of music it is."

You may not have a lot of distinct memories from your childhood; I certainly don't. But I bet you remember the songs, from the biggest Top 40 hits to the dumbest advertising jingles. Calcified by endless replay, the tunes, the lyrics, even the individual piano trills and vocal quirks are probably as

"Well, I took a walk around the world to ease my troubled mind

I left my body lying somewhere in the sands of time

But I watched the world float to the dark side of the moon."

3 DOORS DOWN "Kryptonite"

vivid today as when you first experienced them. Of course, the particular songs that bring *you* back, and how you feel about them, are uniquely yours. But music's remarkable ability to bring our pasts to life belongs to all of us.

And the past is only one available escape route; music can take you anywhere. The immersive experience of music drenches your brain and drowns out all competing inputs, including your sense of the passage of time, the unwritten rules of propriety, the shackles of responsibility. When you're immersed in your music, you're in an isolation tank of thumping bass, soulful guitar, and beautiful, clever, evocative lyrics. Wherever you are, the music you love can take you somewhere else. And it's as easy as turning up the volume.

In this chapter, we'll explore why familiar music takes us back in time, how it helps people escape (alone or in tandem with mind-altering substances), and how it can even help you forget, which can be especially important to certain vulnerable people who badly need that superpower.

## TAKE ME BACK

Here's a superquick exercise, if you'll indulge me: Just say the alphabet in your head, from start to finish. No reading ahead; no cheating. I'll wait. When you're ready, read on.

I invited you to say it. But did you sing it?

As your kindergarten teacher was well aware, music has an almost magical ability to make new memories stick—think of *Schoolhouse Rock*. But it's true at any age—just look at how the unlikely musical *Hamilton* swept the nation, making American history oddly memorable through rap-style lyrics. Anytime we learn something new, whether it's our ABCs or differential calculus, new synaptic chains are formed between sets of neurons in our brain. But these initial chains are weak, easily overwritten by the next shiny experience.

# A DAY IN THE MUSICAL LIFE OF YOUR BRAIN

*You don't have to play music every moment of the day—but then again, it can't hurt. Here are some of the many ways music can be used strategically to enhance every chapter of your day.*

## 2 EXERCISE

*Any good sweat session fights stress, anxiety, and depression, but music gives you extra oomph—science says it can raise endurance by 15%. So pipe something peppy into your ears, and go that extra mile.*

## 1 THE WAKEUP

*It's not easy emerging from your pre-coffee coma. You may assume a harsh alarm is the only way, but choose music instead and your energy will gradually rise to match the song's, reducing grogginess and increasing alertness for the day to come.*

## 6 LUNCHTIME

*Everyone can use a midday brain vacay. Something with a gentle beat, like feel-good folk music, can dial down brain activity and refresh your mind before heading back into the fray.*

## 7 COMMUTING HOME

*To arrive home relaxed even if you left work stressed, don't try to get there all at once. Start with fast-paced music that meets you where you are, then gradually downshift to something more downbeat. You'll hit the driveway ready for the evening.*

## 8 DINNERTIME

*A nutritious meal will nourish your body, but adding the right music to the menu feeds family bonds as well. Familiar music can evoke feelings of nostalgia, boosting oxytocin and bringing the fam just a little closer.*

## 3 SHOWER SHOWTIME

*Belting out a guilty pleasure in the shower (looking at you, "Wake Me Up Before You Go-Go") may earn you a visit from the Noise Police, but showers are prime spots for creativity. Being alone with your thoughts—and the alpha brain waves induced by active tunes—can trigger productive "Eureka!" moments.*

## 4 COMMUTING TO WORK

*Stress can make you white-knuckle the steering wheel—but it can be tamed by a good old dopamine boost. Listening to music you enjoy can increase dopamine levels by up to 9%, helping you leave road rage in the rearview.*

## 5 ON THE JOB

*The right playlists can help you maintain peak work performance, whether you're focusing on a task, calming your nerves before a presentation, or drawing creativity out of your team.*

## 10 SLEEP, GLORIOUS SLEEP

*Lullabies aren't just for babies: Even grown-ass adults can benefit from soothing sounds before bed. Music helps regulate cortisol and promote a blissful slumber. Soundtrack the last 45 minutes of your day with calming tunes to end on a good note.*

## 9 SEXY FUN TIME

*A playlist of songs that make you feel good about yourself can take you from romance right to the main attraction. Familiar love songs will stoke the mood, then thumping, rhythmic R&B will take that mood to its logical conclusion.*

## Why do we daydream?

*Our brain is home to something called the default mode network, a handy system that helps us reflect on inner experiences and develop a sense of self. The default mode network links your high-cognition frontal cortex to your gut-reaction limbic system, as well as to several other areas involved in the sensory experience. Among its many jobs is to help us look back and reflect on our lives, to spin our experiences into self-awareness and a sense of identity. This daydreaming function has a purpose—an analysis of past mistakes might help prevent them from happening again in the future. Today, you can think of it as a creativity engine: Research shows that people whose default mode network is more active are more likely to have clever ideas.*

So how do you make new knowledge stick? The brute-force method is repetition, as when you practice a speech, or repeat somebody's name the first time you meet them, or use flash cards to memorize your times table.

But music has a backdoor hack that gets you there quicker. A song engages multiple parts of your brain—the incoming sound wave lights up your auditory cortex, the rhythm perks up your motor center, the lyrics engage the language center, and the emotional content primes the limbic system—giving the memory a broader footprint inside your head (if that metaphor isn't too convoluted).

Now, when you can combine the two—the broad brain engagement of music and the groove-deepening power of repetition—you can harden that initial synaptic chain into a memory that's almost physically unbreakable. Advertising jingles are a perfect example of this: those short, easy ditties you've heard 10 billion times. I am reasonably certain that in the final fog of my deathbed, the last thing I'll remember, and croak out to my gathered grandchildren in song, is that frosted Lucky Charms are magically delicious. "I was trying to turn times tables into songs, because they're really hard to remember and it shouldn't be hard," says Peter Gabriel. "I think it is another form of glue that could help information stick to the memory, by laying it into music. I still don't think that's been properly done."

A song's ability to transport us to another place or time isn't limited to the experience of hearing the song. This is where **associative memory** comes in: your ability to bind related ideas and store them together. Attached emotions, for example, can signal to the brain that an otherwise nondescript memory is worth preserving, and when you later recall the memory, the emotions are likely to play out again as well, because they're part of the experience now. Your father's favorite song may be a matter of only passing interest to you. But after he passes away, that song might bring you to tears every time you hear it, because that emotion is attached to the memory for you now—it's changed the song for you. Says Citizen Cope: "People have an emotional connection not just to the music, but for that time in their lives. I think we're connecting more to the experiences we had—during those times, with those people—than to the songs themselves."

The neurological mechanisms behind associative memory are complicated, but we know a young brain's ability to build complex, episodic memories combining a who, what, and where goes through a developmental spurt between the ages of eight and 10—lining up, for me, in that "Afternoon Delight" moment. "Nostalgia is the quintessential emotion of memory," says Dr. Petr Janata, cognitive neuroscientist and professor of psychology at the University of California, Davis. "Throughout our lives we often go through concentrated periods of hearing and listening to particular songs, artists, genres, etc., often in social contexts, and so the music ends up being part of the

Familiar music is a comfort in trying times. Streaming services saw a huge increase in nostalgia-themed playlists created in the first week of April 2020.

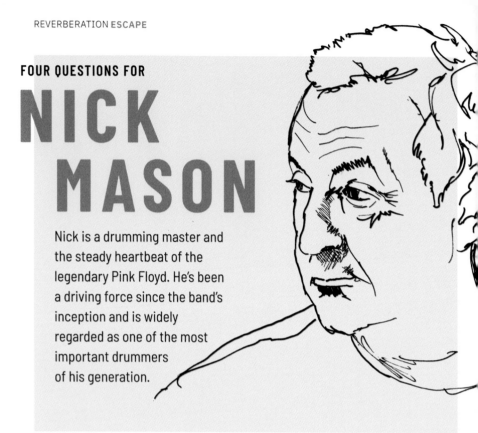

# NICK MASON

Nick is a drumming master and the steady heartbeat of the legendary Pink Floyd. He's been a driving force since the band's inception and is widely regarded as one of the most important drummers of his generation.

**Do you use music to do different things better?**

I use it much more sparingly than I think a lot of other people. If there's one thing that drives me mad, it's people who put music on when they're eating, which sounds sort of slightly cranky, really. But I do find it sort of devalues music to what we always used to complain about, music in restaurants and shops and so on.

I suppose what it really is is that I like music when you're listening to it properly, I suppose, because it's too easy for it to actually interfere with speech. If you're actually talking to someone and have this going on in the background, sometimes it's just really irritating because you're thinking, What's this?

What music is really good for is the drudgery of driving. Music when driving is nice, and it can really shorten the journey incredibly. If the car is an exciting car, it's a distraction to have music. And it's probably going to be interfered with by either motor noise or tire noise.

I've been in a traffic jam and looked in the mirror and seen some, it's always a bloke actually, hitting the steering wheel with drumsticks, but I could actually see them and I thought, I'll turn around.

**Have your musical tastes changed over the years?**

Absolutely not, I still listen to exactly the same things I listened to about 50 years ago. I think it's sort of inevitable that you really get stuck with the music that really influenced you when you were younger. You know, it was really people like John Peele who were extraordinary, he sort of moved with the music and the time. Whereas I'm still stuck with Jimi Hendrix and Eric and Joni Mitchell and all that whole period of people that I knew. I like quite a lot of jazz, particularly, I love Miles Davis. My desert island picks would always include Jack Johnson, the Miles Davis sort of groove school moments, I love that. What I don't do is listen to very much new music or young, young music. Unless you've got someone saying, "You really ought to hear this," you just go, "Let's put Toto on" or whatever.

**Dark Side of the Moon and its unusual sound and instrumentation unleashed a most unusual emotional journey. Was that the plan?**

No, I think it wasn't that premeditated. The initial concept was to have a theme that *Dark Side* was about different areas that we felt bothered us, whether it was mortality or travel, whatever it was. And I think the way of actually assembling the thing came late, and certainly things like the voices. I mean, the instrumentation on *Dark Side* is not particularly interesting, really. There's some sound effects, but possibly more interesting are the voices and what they say. But I don't think we'd originally planned the concept of putting it all together. I think we probably worked out ways of putting some of the songs together, and then it seemed obvious that actually what we should do is make it into one soundscape.

**I'm not driving to *Dark Side*. It doesn't feel right. Do you?**

Yeah. I mean, in many ways it's a sort of headset experience. But the problem is it's very difficult to talk about your own music anyway, because if I listen to *Dark Side*, I don't think, "God I'm brilliant." I think, "We really should have put 'On the Run' somewhere else, it should have come later because it would work better live. Actually, if that little sequence was moved further, deeper into the record"...it's very difficult to listen to your own music without critically thinking.

experiential fabric that characterizes those lifetime periods."

Music's stickiness in your brain helps make it a powerful generator of meaning. We have a specific song for when a president enters a room, for when graduates file up to receive their diplomas, and for when a bride walks down the aisle, to assign these mortal individuals to their proper place in a timeless ceremony with a multigenerational legacy. The songs in your past have that same eternal quality. These are not frozen relics—they're alive; they sink their hooks in all over your brain and pull you through an experience all over again. "Hymns and nursery rhymes have a deep sort of poignancy because these melodies came out of their childhood, and everyone remembers their childhood," says rock & roll singer-songwriter Adam Masterson. "One of the great things about becoming a dad was when I'd hear these songs from childhood again—however corny they might appear—they had this echo of something precious, special, important."

You don't play your music—your music plays you. And while you're in the throes of it, you're simply not in the here and now—you're somewhere else.

## TAKE ME AWAY

How does this happen? The experience of listening to a song you know and love floods so much of your brain that it dampens other inputs. As creatures quite dependent on sensory information, this has the effect of pulling us temporarily out of the flow of our life. The phrase for this is perfect: you "tune out" the world around you. Doesn't matter if you're alone on the highway belting out a classic like "Bohemian Rhapsody" or surrounded by strangers at a concert, entranced and blissfully waving smartphone flashlights in unison. You forget

## I WANT A NEW DRUG

Music and drugs have been bosom buddies for a long, long time. The American Addiction Centers surveyed about 1,000 concertgoers and reported that 57% of respondents admitted they used drugs or alcohol at the venues. Alcohol was the most common intoxicant across the board, and EDM/rave audiences were

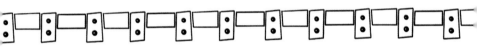

your bills and errands, what time it is and what city you're in, the loves of your life and the challenges of the day. You're content to simply *be*, in this moment, lost in this song.

Escape can be difficult in a world of distraction. Some of us try to clear our heads with meditation or exercise, or by taking that long weekend now and then, though even planning how to spend downtime can itself be stressful. But nothing, nothing lets us step out of the room like music. "When a musical scene draws us in through its rhythmic fabric created by interacting instruments, it provides a rich playground for our brain's attentional and expectation systems," says Janata. In other words, it occupies your whole head—the perfect distraction.

the most intoxicated of the respondents, with two-thirds reporting being drunk or high, with some combination of alcohol, marijuana, ecstasy, hallucinogens, and cocaine, at the event in question. Their number-one explanation (77% of users): To "increase enjoyment."

Yes, I'm well aware there was not a single surprising fact in that paragraph. But consider this: A concert is only one kind of aesthetic experience people enjoy; there are many others. Yet people typically don't get drunk before going to the movies, or take hallucinogens for an art show, or get stoned to increase their enjoyment of a good book. What's different about the concert experience—do drugs and alcohol have a special relationship with music?

## WHISKEY RIVER

Does alcohol make music better? Yes, unquestionably. First, alcohol blurs your perception and impairs your judgment, so you're likely to be less critical. Second, alcohol lowers your inhibitions, so you may make emotional connections you ordinarily wouldn't, resulting in songs feeling more profound, or more personal. Third, alcohol releases chemicals that slow your thinking, increase your pleasure, and relax you, putting you in a receptive mood in which you can pay closer attention to a song.

Whether or not you're drinking specifically to improve your musical experience, music makes you drink more. Yes, music and alcohol are absolutely in cahoots: Studies have shown that the music playing at the bar has a direct effect on how much, and how quickly, you drink, with louder and faster music driving increased consumption. One theory is that loud, upbeat music boosts the brain's arousal generally, prompting a desire for alcohol and the activity of drinking. Another theory: Loud music hampers conversation, causing a bar's patrons to talk less and drink more. (This can be a vicious cycle: Alcohol use partly blocks auditory receptors in the same range as human speech, which is one reason a roomful of drunk people keeps getting louder and louder.)

But there's a third theory as to why music makes you drink more: The presence of music actually alters your perceptions of smell and taste. In one study, a professor

**FUN FACT**

Four in every 10 of Billboard's Hot Country Hits from the previous five years reference drinking...far more than any other musical genre.

144

served students a mixed vodka drink and had them listen to either music, the news, or complete silence. The drinks were perceived to taste sweeter when the students were listening to music. And the music also made it harder to detect how much alcohol was in each serving, an effect that prompted people to drink more than usual.

## EVERYBODY MUST GET STONED

Likewise, marijuana is widely understood to improve the experience of music, giving listeners an almost mystical feeling of being "in the music." Like alcohol, marijuana releases dopamine, increasing your pleasure, and both music and marijuana stimulate your brain's emotional centers, potentially adding an extra layer of meaning to lyrics.

But perhaps most important, marijuana blurs your perception of the normal continuum of time. In his book *The World in Six Songs*, Daniel Levitin explains that marijuana's active ingredient, THC, not only activates pleasure centers, but also disrupts **short-term memory**—and it's this combination that explains how songs take on new profundity when listened to while you're high. "The disruption of short-term memory thrusts listeners into the moment of the music as it unfolds," writes Levitin. "Unable to explicitly keep in mind what has just been played, or to think ahead to what might be played, people stoned on pot tend to hear music from note to note."

**FUN FACT**

Of Spotify's more than 40,000 4/20-themed playlists, Cypress Hill's "Hits From the Bong" is the most streamed track.

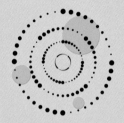

## Why are there so many songs about drinking?

*Because songwriters know you'll love them. And the reason you love them is because of a peculiar synergistic relationship between music and alcohol, wherein each induces consumption of the other. Loud, up-tempo music makes you drink more, for example— and more quickly, too. Interestingly, though, loud, up-tempo songs about drinking can multiply the effect (think of "Shots" by LMFAO, or "One Bourbon, One Scotch, One Beer" by George Thorogood and the Destroyers vibrating through your chest at a club or a pub). In a study published in the American Journal on Addictions, researchers asked participants to listen to similarly styled songs, some with lyrics about alcohol, some without. "Customers who were exposed to music with textual references to alcohol spent significantly more on alcoholic drinks compared to customers in the control condition," the study soberly concluded.*

*So if you're trying to cut back on drinking, perhaps you should avoid tunes that remind you of liquor. And the genre that mentions alcohol the most? Why yes, it is country. That was a lucky guess.*

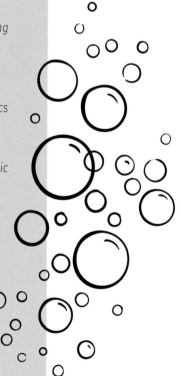

> "I felt like I went to heaven because it was so intense. The music brought me to a place that I had never, ever experienced."
>
> SHEILA E. *on a musical performance providing a religious awakening*

## LUCY IN THE SKY WITH DIAMONDS

Psychedelic drugs, where the experience is so clearly a journey away from reality that we literally call it "tripping," can also heighten the escapism of music. During a 2015 study, researchers asked participants to listen to music while high on LSD, and those who took the drug claimed that the psychedelics enhanced the emotions evoked by the music. And the music makes the acid trip better in return: Listening to music while on LSD increases the amount of information traveling between the parahippocampus and the visual cortex, reportedly causing even trippier hallucinations.

If you notice your hands melting, your face looking unfamiliar in the mirror, or other signs your acid trip is starting to take a bad turn, you need to break that trippier-and-trippier cycle—and since there ain't no turning off the LSD experience once it begins, you might want to turn the music off, or head away from the stage and into the medical tent.

Today, the medical benefits of psychedelics are becoming clearer, and researchers are exploring the promise of carefully controlled doses of music and drugs to improve people's well-being across a wide spectrum of challenges, from curing depression to ending addiction, as Joy Allen reports. "The government was studying LSD in the '60s out in Maryland, and music therapist Helen Bonny was hired to develop music programs to coincide with the stages of LSD effects from beginning to end," she recalls. "We now know those same programs and music can be used to have the same effects of altered states without the LSD, and can be used therapeutically to help individuals process traumas and overcome addictions."

Alcohol, marijuana, and LSD are representative examples, but music has similar symbiotic relationships with other party drugs, and multiple studies are underway to evaluate other drugs for untapped

FOUR QUESTIONS FOR

# CITIZEN COPE

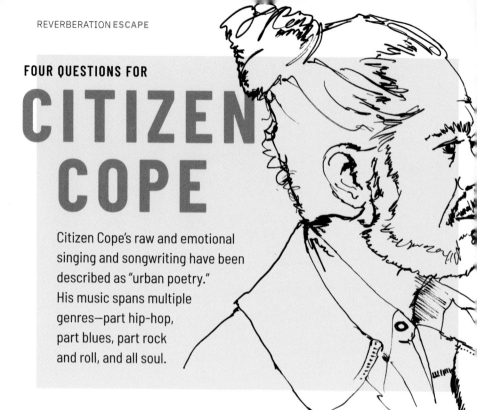

Citizen Cope's raw and emotional singing and songwriting have been described as "urban poetry." His music spans multiple genres—part hip-hop, part blues, part rock and roll, and all soul.

**Is there a particular thing you do that feels different unless there's music on?**

I used to sleep to music when I was younger. I think once I began making records, listening on that kind of level is different. Recently, I kind of went through a period where I wasn't listening to as much music, and now I'm listening to more again. It's really amazing how deep and how powerful it is, how much you can learn from it.

But I think, honestly, when you start making records, as a producer, the enjoyment can be obstructed by trying to pick apart a song or analyze it more as a producer and not as a listener. So I think there is that fine line of when you start making records and when you start listening to records. That naiveté that goes into it, that ideal as a listener, is something that I'm kind of working on right now.

**When you're writing a song, is the process more intellectual or more emotional?**

I think it's definitely more spiritual for me. It's kind of like a reaction. It's like a breath in and a breath out, you know, like something hopefully instinctual. It kind of led me personally as a songwriter. I never thought it would. I guess it was just...it was something that answered some questions that I wanted to be answered personally, spiritually. I guess that was the vehicle that got me there.

Sometimes I'd be kind of feeling off or a little sad or something. Some of the songs somehow explained scenarios in my life, maybe things that happened that were bad. They gave them a reason. The songs were able to take me out of a depression, I think, early on, hard times. I had some losses in my family, so I just started playing around with this guitar and despite knowing that I had a good sense of writing, I wasn't particularly musically gifted. But the songs came out.

I love listening to music and just started following that muse. And I think that music carries you to these beautiful places.

**Can you describe the moment you step onstage and there's peak energy in the room, and fans are chanting your name?**

I don't know. I'm not the best person to ask about it, 'cause I'd been intimidated by the audience for so long. I look at this as a higher purpose, you know. I look at music as a way to reach. It was kind of a stepping stone for my own personal spiritual kind of enlightenment, and also an expression and possibility of that, as opposed to individual fame and fortune and stuff. Not that Michael Jackson and Prince were looking at it from only fame and fortune, but I think there was a big change in the industry when those real artists became these huge, iconic pop stars...

It just never was about my own personal thing. It was about something higher. Even though you kind of lose sight of that because you have to sell records—which is really a mind screw—and you end up asking yourself, what master are you serving?

**Have you gone out onstage when you've been, let's call it, less than sober? Is there a difference in going out onstage and performing when you're sober, as opposed to possibly not so sober?**

Well, that's a deep one, because I struggle with drinking too much. And I think I had such bad stage fright, I used to drink a little bit before. Then I went through periods where I didn't drink during the show, but afterwards. So then, I don't know if I'm hung over the next day while I'm playing the show, as opposed to drunk, and, yeah, there's definitely a different vibe.

I think this music thing, and songwriting, to me is just a path to personal growth. And I, like—at first I thought it was about the spiritual thing. Then I was like, okay, what can I personally attain through this? And that's where it gets confusing. You go from autonomy to being praised, which is also a weird thing. And being recognized for something. But I think it comes from a higher power. I don't know the name of that higher power or whatever it is, but there's something pretty mystical about music, I can't put my hand on it.

therapeutic potential. Maybe someday in the not-too-distant future, your doctor might write you a prescription for a drop of acid and a healthful playlist. "Take two Jimi Hendrix deep cuts and call me if your symptoms persist."

## I WANNA SEE YOU BE BRAVE

On the darker side of the escape from reality is trauma, including post-traumatic stress disorder, or PTSD, which represents a monumental challenge for mental health profession. Often characterized by persistent unwanted memories of past events (experienced personally or witnessed), trauma is a condition whose destructive re-triggering cycle can be difficult to short-circuit or counteract, making recovery for sufferers painful, prolonged, and unpredictable. Music, as a uniquely powerful tool to *strengthen* memories, might seem an unlikely candidate to help trauma sufferers keep unwanted memories at bay. But music's versatility runs deep.

Trauma's long-term effects can stem from childhood neglect, sexual abuse, a horrific accident, or domestic violence, to name just a few. It can be acute (caused by a single event), chronic, or even vicarious (e.g., initiated by watching a loved one suffer). And trauma can be triggered again and again, even by innocuous stimuli.

When a triggering event occurs, a number of things happen in the trauma sufferer's brain—let's call her Sally. Activity is suppressed in Sally's emotion-and-impulse regulator (part of the prefrontal cortex), leaving her less able to control feelings of fear and shifting her toward a purely reactive state. Her memory control center (the hippocampus) starts blending the past and present, sowing confusion as to whether the trigger is itself a traumatic event. And her fear-response generators (the amygdalae) might go into overdrive, causing her to feel very much like she's reliving the traumatic event itself, rather than simply remembering it. Over and over again.

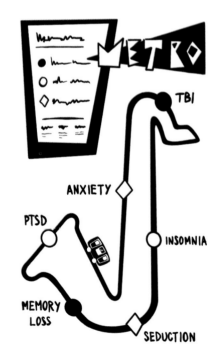

# The **Doorway** Effect

*The dumbest advertising jingles are burned into your brain forever. But run into the house to grab your car keys and you can find yourself standing in a room with your hands on your hips, trying to remember what you came in for. Why the difference? Because these two memories aren't stored in the same way in your brain. That jingle from 30 years ago is encoded into multiple regions: your motor cortex, your auditory center, your limbic system, your language center. The location of those misplaced car keys, on the other hand, is just a short-term memory phenomenon, and that's why you fall victim to the "doorway effect." (Cue the spooky music...)*

*You know the feeling: You're on a simple errand, step into the room, and promptly forget why you came in. Believe it or not, it's the doorway that's the problem. We have a natural instinct, no doubt honed by evolution, to reassess everything when we enter a new room. Are there new threats here, like that sketchy-looking cat? New opportunities, like a bowl of M&Ms on the piano? Entering a room changes your environment, and that requires at least a subconscious reassessment, which competes for our working memory's limited resources. And that's why you're standing there like an idiot, with the keys still hanging forlorn on the hook.*

With all those structural changes to the brain, it's perhaps no wonder that trauma can cause insomnia and fatigue, increase the risk of addiction, and lead to mood disorders such as anxiety and depression that are difficult to walk back and can cause terrible suffering: People with PTSD are two to five times more likely to die by suicide.

Music therapy isn't the whole answer to the complexity of trauma. But it can be an important and valuable tool in ameliorating symptoms and improving sufferers' well-being. As a complement to traditional cognitive or behavioral therapy,

music therapy can help trauma sufferers build coping strategies that recalibrate their behavioral responses and foster a sense of safety and security.

But there are dangers, and music therapists have to tread lightly. "When we're using it with PTSD, it's to bring us to a place where we feel safe and secure, where we're not feeling threatened," says Joy Allen. "Music can be used as an escape, which is a great thing to self-soothe and deactivate. But at the same time we have to remember that it can trigger—we have to be careful." Because music has such a powerful ability to resurrect old memories, a music therapist could inadvertently trigger the negative response they're trying to prevent if they choose the wrong music.

German philosopher Friedrich Nietzsche once said that a memory from the past sometimes "returns as a ghost and disturbs the peace of a later moment." That's a pretty good working definition of the mechanism behind PTSD. In World War I, the condition was called "shell shock," assumed (wrongly) to be brain damage caused by proximity to exploding mortar shells. By World War II it had become "battle fatigue," and we had at least begun to understand it as its own distinct psychological condition. It wasn't until 1980 that the condition was formally defined as PTSD, a designation that's still being refined.

**FUN FACT**

The **iso principle** is a technique by which music is matched with the mood of a client, then gradually altered to affect the desired mood state.

The U.S. Department of Veterans Affairs, which has an obvious ongoing interest in improving outcomes for PTSD sufferers, once commissioned a broad-based theoretical review of PTSD-related studies. The review found that music therapy improved social functioning in four specific areas: community building, emotion regulation, increased pleasure, and anxiety reduction. As the authors put it, "Music therapy may be considered a resilience-enhancing intervention as it can help trauma-exposed individuals harness their ability to recover elements of normality in their life following great adversity."

One quick example of music therapy in action: Six soldiers suffering from PTSD were invited to join a therapeutic drumming group. The soldiers reported an "increased sense of openness, togetherness, belonging, sharing, closeness, connectedness and intimacy, as well as achieving a non-intimidating access to traumatic memories, facilitating an outlet for rage and regaining a sense of self-control."

Memories degrade naturally over time, and it may seem intuitive that forgetting is simply the end state of that natural process. Not so: It turns out that forgetting is a definitive act, something that your brain chooses to do. There exist mental mechanisms that promote what we now call "active forgetting," and for conditions like PTSD and drug addiction, where people may be trying to leave something behind, these are considered promising therapies. ◼

# TAKEAWAYS

Music is so much more than a pleasant distraction—it's a whole-brain experience that temporarily transports you out of the real world. If you've ever sat, eyes-closed in your car cranking the end of a song before getting out to fulfill an errand, you know what I'm talking about. Songs are tightly connected to memory—in fact, they're exactly the kinds of patterned experiences our memories hunger for—and they can stubbornly persist in our heads, whether we want them to or not. The right song can take you to another place or time, can help you escape the daily grind without the side effects of drugs or alcohol, and can even help mitigate the disturbing effects of unwanted memories, including traumatic experiences.

**TO MEMORIZE SOMETHING IMPORTANT** Try coding it into a musical jingle and singing it to yourself. Our brain assumes songs are valuable, coding them in multiple places for more reliable retention later.

**TO HELP A FRIEND DEAL WITH TRAUMA** Suggest a credentialed music therapist as a complement to her or his conventional treatment. The right approach can really make a difference.

**TO NEUTRALIZE SUBCONSCIOUS TRIGGERS THAT CAUSE YOU TO DRINK TO EXCESS** Find bars that keep the music a little slower and a little less loud, so you're not speeding up your drinking to compensate for these environmental factors. At home, keep songs about drinking out of your playlist—minimize the inputs your subconscious mind gets about drinking and it can slow your consumption.

CREATE

# I'M TRYIN'
# SOUL WITH
# GONNA
# THE REST OF

**BOB DYLAN**
"Working Man's
 Blues"

# TO FEED MY THOUGHT

# SLEEP OFF THE DAY

**IT WAS THE FALL OF 1762**, and the famous pianist Wolfgang Amadeus Mozart was about to play a command performance for the Austrian emperor. Before sitting down, Mozart asked the emperor whether the court composer was in attendance, as he was about to play one of the man's compositions. At the emperor's bidding, the composer emerged from the wings and was introduced to Mozart—who promptly asked this famous man to *turn the pages for him while he played*. It was a potentially awkward moment, but Mozart was easily forgiven.

After all, he was only seven years old.

Mozart is possibly the most famous child prodigy in history. Squired around by his family, he toured the royal courts and concert halls of Europe, flawlessly playing the most difficult piano pieces set in front of him. He reportedly taught himself to play the violin, the organ, and the harpsichord. He inspired Beethoven; he wrote an opera for Archduke Ferdinand's wedding; he was knighted by the Pope.

## Do animals respond to music?

*It's been noted that dogs pay attention to dancing, but they might just be reacting to humans gyrating strangely. By and large—except for birds—animals don't seem to pay much attention to music...not music composed by humans for humans, anyway. But animal psychologist (why yes, it is a thing) Charles Snowdon has done pioneering work creating music with pitches, tones, and tempos matching those employed by cats and monkeys when they communicate, and has reportedly enticed them to pay attention. We're not quite sure meow.*

But it wasn't just about talent: Mozart was an unparalleled creative force. He was a prodigious composer who created more than 600 masterful works—including 30 symphonies before his 18th birthday—across a breathtaking range of styles. He wrote offertories for Benedictine monks and sonatas for the Queen of England; he composed great operas, choral works, chamber music, string quartets and quintets, and even comical pieces designed specifically to sound clumsy and make audiences laugh. His brain just seemed to operate differently than the brains of us shabby mortals. One time, in the Sistine Chapel at age 14, he heard a performance of the famous *Miserere*—a complex, five-part interlocking choral piece whose score was a closely guarded Vatican secret. Later that afternoon, alone in his room, the teenager wrote it all out from memory. End of secret.

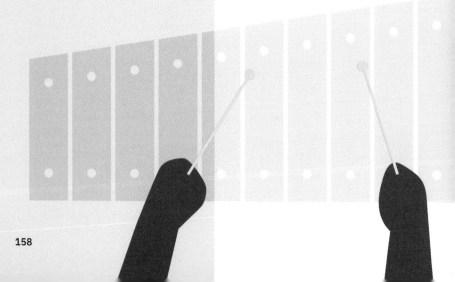

The mystique around child prodigies, the magical thinking, is that this sort of creativity is a gift delivered straight from heaven to a lucky few at birth. But the reality is more prosaic. For the most part, prodigies—musicians, artists, athletes, chess players—are smart, talented, driven young people who practice a lot. And that's all.

Musical prodigies, in particular, have been studied extensively, and a few things seem to set them apart from the merely talented:

1) They have an external motivation to begin playing—often a pushy parent.

2) They share a propensity for a very high level of practice, beginning at an early age.

3) They have a higher-than-average tendency to experience flow states during practice (for more on flow states, aka "getting in the zone," see Chapter 2: Focus).

There may be other commonalities: One study, for example, found that art, math, and musical prodigies typically have superior working memory. But generally speaking, child prodigies are made, not born: They're motivated by something or someone, they practice like demons, and they get deeply, thoroughly immersed in their work. In an article for *Scientific American*, psychology professor David Z. Hambrick declares that most child prodigies share what he calls a "rage to master," and pour an almost unfathomable amount of time and effort into their area of expertise.

Man on New York street corner: "How do you get to Carnegie Hall?"

Wise neurologist: "Practice."

Creativity isn't a special trait held by a fortunate few—it is your birthright, as a human. Humans may not all be geniuses, as a trip to any Department of Motor Vehicles will quickly remind you. But we are all creative; this is an adaptive trait we share, and part and parcel of what it means to be human. Our cushy modern life doesn't always demand much creativity, so for many of us it's easy to dismiss creativity as something only artsy people have. But that's a shame, because if you can tap into your natural creativity, in any field, you have a shot at changing the rules—and thereby changing everything.

# WHAT CREATIVITY IS

To understand how music can be leveraged for creativity, I spoke with Dr. Charles Limb, one of the foremost researchers into the neural basis of musical creativity, and Stanford professor and sleep specialist Dr. Rafael Pelayo, who both encouraged me to think about creativity not in today's narrower "fun storytelling" definition but in the broadest possible sense: as a technique for solving problems. "Problem-solving is just part of being an animal," says Pelayo. "Whether you're an ant or a human, you have to be able to problem-solve in real time." As the world changes, creatures with the ability to find novel solutions to emerging challenges outcompete their peers, generation after generation. Creativity isn't a curious sideshow; it's a core survival mechanism.

Limb agrees, putting it succinctly: "The capacity for problem-solving and innovation is a cornerstone of human survival." Our brains are literally wired to look for patterns; this is how we understand and organize the world into truths that help us survive, such as "Plants like this may be safely eaten," or "This lake disappears in winter, but reappears every spring." Pattern recognition helps us accumulate knowledge we can communicate to others; creativity is what's called for when the pattern is broken. When the lake *doesn't* reappear as expected in spring, and the whole tribe is

# In 2016, Mozart sold more CDs than Beyoncé. (Thanks to a 25-pound, 250-CD boxed set that only had to sell 6,250 copies.)

thirsty and confused, creativity equips us to do something about it.

## SO HOW DOES IT WORK?

Very generally speaking, you solve creative challenges in a two-stage process. First you review everything relevant: the patterns you know. Then you try to alter your mindset somehow, reviewing the same information from a different angle in search of new patterns that can lead you to new answers. The facts don't change—it's your perspective that changes. Creative breakthroughs are transcendent: brilliant in retrospect, but unexpected in the moment. A great quote attributed (dubiously) to automobile pioneer Henry Ford goes, "If I'd asked people what they wanted, they'd have said faster horses."

Whether you're a basketball player who's creative on the court, a deejay improvising in a club, or a head of product marketing looking for a new approach to beat the competition, we are all challenged continuously to adapt to changing environments. That's what the human experience has always been—and creativity is the tool that reliably gets us to the other side.

## HOW TO TAP INTO YOUR CREATIVITY

You may be surprised to learn that Paul McCartney didn't write the Beatles' classic "Yesterday"—the song actually wrote itself. By his own account, the cute Beatle woke up one morning with the tune already in his head, having come to him in a dream. It was so complete that it felt like a familiar song, and he spent the next few weeks playing it for everyone he knew, asking if they'd heard it before, before he felt confident enough to claim it as his own.

When your goal is to think differently—to create game-changing pop songs, or to solve thorny business challenges—there are many ways to alter your traditional mindset, drugs and alcohol being common favorites. But another natural mechanism that often can serve this purpose—the one Paul unconsciously applied—is sleep.

Turns out we go through a nontrivial mindset shift each and every night, as our brain switches off our consciousness and gets to work consolidating another day of reality for us, laying down another ring in the tree trunk of our lives. This seems to consist of replaying the inputs of the day, comparing them against existing patterns, inferring new patterns, and cataloging what seems to have happened and why it matters. Recent research from UCLA's Gina Poe and colleagues suggests this happens in two phases: In non-REM sleep, we code the day's memories and experiences and come to a general understanding of the gist of the day's events, and in REM sleep, we form new and unexpected connections and consolidate these into new ideas.

The idea that sleep provides a natural reservoir of creative inspiration is not new, and about 100 years ago there was a vogue for trying to leverage this natural altered state for creative inspiration. Surrealist painter Salvador Dalí, philosopher Jean Paul Sartre, moody proto-goth Edgar Allan Poe, and manic inventor Thomas Edison were just a few of the proponents of exploring "hypnagogia," the transitional state you know as falling asleep, for its inspirational possibilities.

During **hypnagogia**, your eyes are closed; you're relaxed; your brain-wave activity is slowing down. Your body may twitch as your motor controls begin to shut down for the night; your mind begins to wander in crazy directions as you transition from the real world to the dream world, from noticing to imagining. This is one of the sleep phases where **lucid dreaming** can happen, and Dalí and his fellow hypnagogianauts tried to capture this phase's

**FUN FACT**

Monaco has fewer soldiers in its army than musicians in its military orchestra— 82 soldiers vs. 85 musicians.

rich creative meandering by snapping themselves out of it, mid-phase, with sleep deprivation.

Dalí, the mustachioed madman who was painting melting clocks and other disturbing dream imagery before LSD was ever synthesized, would rouse himself from hypnagogia in the following way: He would set a plate upside down on the ground next to a comfortable chair, then sit down holding or balancing a small metal object, like a spoon, above the plate. As he started to drift off to sleep, his motor control would eventually slip, letting the spoon clatter onto the plate, jarring him awake—whereupon he'd instantly record his thoughts and dream imagery, however bizarre. Edison did the same thing, almost-napping in his office chair multiple times a day, dropping a steel ball onto an overturned metal plate on the floor.

Targeted sleep intervention may be just the drug to spark your creative insight—it's hard to argue with "The Raven" and the light bulb. But music may provide a more reliable path to creative greatness.

## HOW MUSIC HELPS

It's easy these days to dismiss music as pure entertainment, an occasional escape or something to play in the background while we do something else. But historically music has been much more mission-critical. It seems to have been around as long as civilization, and possibly is the precursor to language itself. However we got here, it's easy to see today, with modern brain scans, that our brains still take music very, very seriously.

"Music is my first love, it's a language that doesn't need to be translated," says world music superstar Youssou N'Dour. "Before the words come in, the music and the

FOUR QUESTIONS FOR

# BRANFORD MARSALIS

Born into a multigenerational family of musicians who have ruled New Orleans and its legendary jazz scene, Branford stands as one of the most beloved musicians of his generation.

**What's in your playlist?**

Well, I don't know. I put whole records in the playlist. I don't have a playlist with a multitude of songs that are supposed to give people some kind of strange impression of the kind of person I am. You know? So I'm listening to *Götterdämmerung*. 'What's that?' Oh, it's an opera by Wagner. 'Well, you know what else is in your playlist?' I'm like, that's it, that's in the playlist. That's what I'm listening to. 'How long have you been listening to that?' Two years.

**Do you feel that you write songs differently depending on your mood?**

I don't write lyrics. So the great thing about instrumental music is it's the ultimate Rorschach test. I mean, whatever people want to say it is, that's what it is. But with lyrics, it's very personal to them.

So I get other people to write the lyrics and then I change them as I feel I need to, for it to make sense to me.

...And it's very different than now, where most kids have these digital audio workstations and the only songs they really listen to are their own. And that's the great disconnect. There's no connection to any past, because these days you can be a 10-year-old writing your own songs.

And if you write enough of them, you might stumble and get two or three good ones or 10 good ones, but there's that lack of connection to the past. That connection to the continuum is missing more than it used to be, in earlier and earlier years.

And I think that that is a huge difference between how young people listen to music, and why they listen to music. There's a means to an end they seek that has very little to do with music. It has more to do with using music to achieve some sort of supercharged success that they envision for themselves. And that's a very, very different way of approaching.

**Do you find a difference in how you're feeling as you're improvising versus playing something rehearsed?**

No, it's pretty much like sports, I guess. You do a rigorous amount of training. And then when the training's done, you go out and you play a game. The less training you have, the less likely you are to succeed. The less you watch film, the less you analyze what's out there. It'd be like sticking me on a football pitch. I'd just be standing up. I wouldn't know the first thing to do. It doesn't mean I don't have any athletic gifts. It's just, I don't know anything. And I think this is the problem with the myth of improvisation. It is based on this mistaken assumption that you can play anything, but what you are playing has to be relative to the people you're playing with, which means that you have to have a tremendous sound palette. If you're going to play with an orchestra and then suddenly play with Sting or play with Dizzy Gillespie, you have to be aware there is no one-sound-fits-all-approach to this thing.

So improvisation to me is like language. I think that's the better analogy. There are conversations I can have with my friends...and that's what improvisation is. There's a reason why the majority of people who like saying they know how to play jazz always play songs that are modal and in a minor key, because the one scale they know works in that setting.

And it's the same complaint that I have about modern jazz musicians, that what you are seeing they're doing is very different from what they're actually doing, because what they are doing is rehearsing. And then they go onstage and they regurgitate the things that they've been rehearsing, which is not really improvisation. You learn all these things, and then you shut the switch off and go out there and you allow it to come to you and you react to it.

**Is playing music always fun?**

When I pick up the instrument, I'm there to work. It's just...it's just a grind. You know, I'm just reading a bunch of notes and trying to learn, reading things that are demanding and difficult in going through the slog. I mean, there's nothing exhilarating about it. It's taxing. It's boring, but it's necessary.

sounds provoke an emotional response, and you can touch someone who doesn't understand your language."

When we hear music, its effects flood through nearly every part of our brain, including areas associated with movement, emotional processing, memory, language, creativity, and more. Dr. Limb told me that his researchers once scanned the brain of a blind man listening to music, and the man's visual cortex lit up like a Christmas tree. "Is there another experience that delivers a similarly comprehensive whole brain explosion? The only one that comes to mind is sex. Music is *important*.

"Music is so multi-dimensional, with different circuits in the brain keeping track of different attributes, like **pitch** and rhythm and loudness and tempo and contour and tonality," says Daniel Levitin. "You've got all these multiple pathways converging to form a musical memory, to the point that even if you only have a few of the features, you can recognize the song."

He tried this quick experiment with me, clapping out this pattern with his hands:

*Clap-clap-clap*

*Clap-clap-clap*

*Clap-clap-clap clap clap*

With no melody, lyrics, or context at all, this beat was instantly recognizable: It's the "to-the-dump, to-the-dump, to-the-dump dump dump" of Rossini's *William Tell Overture*. (If this demo didn't work

## Why do I have my best creative ideas in the shower?

*Because you're in a relaxed mindset, doing a simple, repetitive task you've done 10,000 times before inside a confined and safe space. It's literally the perfect recipe for putting to good use your mind's natural tendency to wander. "When are you alone with your thoughts?" wonders Rafael Pelayo. "When you're in the shower, and when you're in bed, waiting to fall asleep."*

for you, it's likely the fault of my notation—trust me, it works in person.)

So why do our brains, which can process only a tiny portion of our total experience, care so much about music? Remember that the brain is always on the lookout for patterns, and songs are all *about* patterns. The seven-part **scale**, the beats per measure, the interplay of **melody** and harmony. The parallel structure of song verses, the repetition of the chorus, the way lyrics rhyme. Even the simplest songs present multiple patterns for decoding, with apparent rules established and then bent or broken. And it's all manna for our pattern-hungry brains.

So it stands to reason that we should be able to leverage music to improve creativity—and indeed we can. The same alpha waves generated by your brain when you're in a relaxed, thoughtful, "waking rest" state are, perhaps unsurprisingly, the ones linked to creativity as well. You can get there through mindfulness and meditation, biofeedback, or your drug or cocktail of choice. Or you can try listening to music in the 50–80 bpm range discussed in Chapter 1: Relax. These alpha waves have specifically been associated with the "aha" or "eureka" moment—that famous flash of insight so momentous you're tempted to announce it out loud.

At one level, music can boost creative focus simply by occupying part of your busy mind, keeping distraction at bay. "As a painter, animator, and digital artist,

I love having music playing while I work, especially during long creation sessions," says Diana Saville, cofounder and COO of brain-industry accelerator BrainMind. "For those sessions, I tend to rely on albums I have listened to hundreds of times: This provides the comfort of familiarity and is guaranteed not to be too distracting. The right music can feel like it's a creative partner, helping you stay in the zone." Her anecdotal evidence is supported by the science: Research by University of Miami professor Teresa Lesiuk and others indicates that when people in creative fields listen to music, they work faster and come up with higher-quality ideas than when they work in silence.

Dr. Heather Berlin, neuroscientist, clinical psychologist, and associate clinical professor of psychiatry at the Icahn School of Medicine at Mount Sinai, has a theory as to why this works: "Creativity and improvisation can be a way to access people's unconscious. When you are in these creative states, you have decreased activation of the prefrontal parts of the brain that normally suppress the amygdala and the hippocampus. And so if you can get people into these states and release that suppression, you can more easily access these more unconscious or suppressed thoughts and emotions."

Another way songs can help jump-start the creative process is by exposing yourself to unfamiliar music—for example, music from a different era, or of a different genre. Challenging your brain to comprehend sound in a new way—to interpret new (to you) patterns—can open the door to new connections. Another potential lever is working with positive music: One study suggests that listening to happy music can create more divergent thinking, a main ingredient in creativity. Volume matters, too: Moderate levels of music (and other ambient noise) have been shown to be ideal for promoting abstract processing and pushing listeners to higher levels of creativity.

But it's good to remember that creativity is tough to control or influence, even for professional creatives. "Being a songwriter is like being a nun: You're married to a mystery," famed composer Leonard Cohen told an audience one night. "If I knew where the good songs came from, I'd go there more often."

## I JUST WANNA BANG ON THE DRUM ALL DAY

If you're not satisfied with hacking your way to a creative breakthrough now and then, and instead want to level up and become a more creative person in general, you might consider picking up a musical instrument. We know anecdotally that music and math have a relationship; the enduring image of Einstein and his violin burned *that*

**FUN FACT**

41% of Americans now use online streaming services as their preferred method of listening to music. That's a huge shift from 2017, when radio was, *by far,* the most popular way.

into everybody's mind a few generations ago. But we have learned a lot more in recent years about how music training can specifically mold the brain into a stronger, more flexible machine more capable of creative feats.

Looking at **functional MRIs** (fMRIs) that visualize brain structure and activity, researchers have discovered that musicians' brains are physically different from those of nonmusicians. In fact, experienced anatomists can look at a few fMRI images and identify which brains belong to musical pros. To be clear, musicians weren't born with these finely tuned instruments—rather, their professional-grade musical brains were laboriously built and continually refined through relentless practice. And the way that practice changes brain structure is through the miracle of **neuroplasticity**.

Neuroplasticity describes the brain's borderline-mystical ability to fundamentally rebuild itself based on its inputs, every minute of every day, over the course of your life. Your brain's physical, chemical, and electrical structure isn't fixed—it's in a constant state of flux. Synaptic pathways are strengthened with repetition, or degrade over time, to gradually be overwritten with other pathways, other memories, other habits. New neural connections are forged; alternate neural pathways are constructed—the road crews in your head never quit. For musicians, long hours of practice strengthen particular synaptic connections, build habits, develop mind-hand coordination, and so on, and all this can be read in the structure of their finely tuned brains, optimized by music.

FOUR QUESTIONS FOR

# SHEILA E.

Sheila E.—also known as "The Queen of Percussion"—has always followed the beat of her own drum. Beyond her own personal successes, the legendary percussionist, drummer, and singer's star shone brightly through her famous collaborations with Marvin Gaye, Lionel Richie, and Prince.

**There's a comparison of a runner's high to a drummer's high. Can you describe the similarity?**

Well, it's really interesting because I was an athlete first, I was a runner. I was training to be in the Olympics for track and field. So the runner's high is all that I knew. And that high that I got from running was everything.

Everything just looks incredibly brilliant, vibrant, after running, even though you're exhausted...

I played music with my dad and I loved music, but one time playing with him, I realized this high was different because it became more like a spiritual high.

That one show that I played with my dad at 15, I really believe I left my body. All I remember was I closed my eyes when the band's playing, he tells me to take a solo. All of a sudden I am looking at myself, looking down at the band and the entire venue with the audience, like, "How am I watching myself play?" I felt like I went to heaven because it was so intense. The music brought me to a place that I had never, ever experienced, even through the things that I had done in sports. So this became a spiritual moment of an out-of-body experience.

And when we got offstage, I grabbed my dad, and I was like, "Daddy, Daddy, this is what I wanna do for the rest of my life. I wanna be a percussionist." And I was crying. Then he started crying. I think I went to heaven, and I wanna feel like that every single day. I know God he's telling me this is a gift. It was emotional. It was bigger than life.

**How does music guide your day?**

Music varies throughout the day, it depends on "All right, I gotta get going," so I need to play some salsa music, I need to dance or, you know, old-school Motown, or it'll be classical music later in the evening. My music changes all throughout the day, depending on how I would like to feel, because music puts me in a place and in a mood of feeling happy, giving me energy. When I get ready to work out, I need music that's gonna give me power to get through the things I won't be able to get through.

If I'm trying to get through something like working out and I really have to push, if I put on some relaxing music it messes with my mind. My body's saying, "Go!" and my mind's listening to this music saying, "Yeah... but I wanna lie down now," that's not the right music. So it does affect what I would listen to. For me, it's how I would like to feel that mood, and definitely it changes in genres of music or nature, things like that.

**What's your surefire writing inspiration?**

I can't always sit and say, "I'm gonna write about this." I could be any-where, it could be in the shower. I could be somewhere driving and I get a word, I think of something or I see something on the street, "Oh, that's a good idea for a song." I mean, I'll just yell it out, "There's a new song!"

**How has music helped raise your kids?**

I should go back to my parents raising us. We had music in the house constantly. So you know, my dad would say, "Okay, you guys, don't touch my instruments," and then he would leave and then we'd play. And then we would be in trouble for playing his instruments. And then he realized that there was something that we could really do together. I'm so thankful that we have music in our family.

The focus of music was, "Let's put on music and let's dance and let's entertain." It seemed like that happened every single day. So when the family functions happened, we were the entertainment. You know, we were the Jackson Five, we were the Osmonds, we were the Tempta-tions, the Supremes, we were James Brown.

So I have to go back to them raising us in a way that was so much fun, to just bring in music every single day. It's just like, we're brushing our teeth, we're drinking water, we're eating, and now we're gonna dance. I mean, who does that? That's just kind of awesome. So trickling that down to all of our kids, everyone, we all do the same thing.

## Will playing classical music give me a genius baby?

*Nope. This popular myth (a whole business was built around it) was based on a misreading of a study which found that listening to classical music helped college students (not babies) learn, and the effect lasted all of 15 minutes. However, scientists are in wide agreement that music—played at a low level; their ears are sensitive— is good for babies' development, particularly building their language skills.*

Here's a snapshot of some of what happens inside the mind of professional musicians. When composing music, musicians strengthen the specific neural networks responsible for processing emotion and integrating information. Performing music, by contrast, is more like a task; it reinforces the same neural pathways activated when you're doing math. This makes sense: You're playing something you've rehearsed, and strictly obeying the rules. By contrast, when artists are improvising—breaking the rules to create something new— activity slows down in the prefrontal cortex, where planning and self-monitoring happens. This loosening of the reins seems to make room for creativity to flourish.

Remember, the point isn't that musicians are special—your brain is just as plastic as Yo-Yo Ma's. In fact, studies have found that the brainwaves of audience members actually sync up with performers' during a performance, which seems downright miraculous. The more you enjoy the musical performance, the more synchronized your brainwaves will be, to the point where the level of this "inter-brain coherence" can actually predict the listener's enjoyment of the performance. If it feels like you're one with the band at a concert, in an important way you really are.

We've long known that musical training is good for the brain; studies show, for example, that children who play musical instruments can perform complex

# BUILD **MORE** POWERFUL BRAINSTORMS

*A productive brainstorming meeting has two parts: an initial, blue-sky, think-outside-the-box **idea-finding phase**, and a process of filtering through these ideas to build out the most promising options—the **solution-finding phase**.*

*To get your team's brainwaves where you want them to be, put together a two-part classical-music playlist. (Classical has been shown to challenge the brain and inspire creativity, and because it's likely to be unfamiliar to your team, it won't be overly distracting.) Try some fugues or chamber music for the first half, to generate the alpha waves perfect for creativity; for the second half speed it up a little with some Vivaldi ("Summer," from The Four Seasons), or something similar, to get busy beta or even ultra-busy gamma wave patterns flowing.*

mathematical problems better than their peers. It doesn't matter whether or not you make it to Juilliard— it's the training that counts. Strengthening your tools for creative expression is important for physical health, providing strong neural pathways that will help you maintain health later in life. The executive director of the International Arts + Mind Lab at Johns Hopkins's Pedersen Brain Science Institute, Susan Magsamen, told me, "Creative expression is at the core of the human experience. If you can't find and share and have your voice celebrated, you're inevitably not going to feel healthy and whole."

Magsamen developed a research model called Impact Thinking that brings the arts, science, technology and health together to better study the impact of the arts in our lives; it is one of many budding efforts around the world that promote creativity as an important exercise in being fully human. "It's not that you need to be a painter who shows their work in galleries, but you need to be able to figure out the story of your life—who you are and how you want to live," says Magsamen. "You may not be a published poet, but you can create meta-phors to give you insight; to learn more about yourself through perspective taking and empathy. These are just great, essential skills for how to be a whole person."

If creativity helped make humans prosper in the past, there's no reason to think we won't need this superpower even more as we face off against a host of seemingly intractable global problems. "This creative capacity, the ability to problem-solve and innovate, is very fun-damental not just to understanding how we became who we are today but to who we will be in the future," says Charles Limb. "You might actually argue from that perspective that art and artists and creativity are more important than any other topic you can imagine."

So please, for the sake of humanity—pick up that banjo. ■

# TAKEAWAYS

*Creativity is your birthright—a naturally evolved ability to solve challenges by thinking outside the box and an important part of what makes you human. Used strategically, by addressing the brainwave patterns known to optimize creativity, music can be a natural, low-cost means to unlocking the potential for innovative problem-solving, no matter what the situation.*

**TO FIND AN ELUSIVE SOLUTION** Listen to unfamiliar music from a different era. If the answer you're looking for can't be found in the familiar way you've been approaching a challenge, take time to tap into a mental space where your mind can free-associate your way to the "aha!" moment. Find a safe place to be alone with your thoughts (the bedroom, for example), and cultivate a meditative, exploring mindset through unfamiliar music. This suggests a new environment to your pattern-hungry brain, arming it to try seeking a novel solution.

**TO INSPIRE A GROUP TO COLLABORATE WELL** Have everyone contribute to a playlist of positive, upbeat music. Happy music has been shown to lead to divergent thinking, and mutual enjoyment of music can sync up the brainwaves of the audience, which can aid in productive collaboration.

**TO GRADUALLY BECOME A MORE CREATIVE PERSON** Try some musical training. You don't have to be young, and you don't have to be good—thanks to brain plasticity, it's never too late to create or reinforce neural pathways that lead to clearer critical reasoning and improve your ability to think outside the box.

# FEEL

your mood
choose
l'avatparation's

# BUT DON'T TELL MY MY ACHY

**BILLY RAY CYRUS**
"Achy Breaky Heart"

# HEART
# BREAKY
# HEART

**I WAS BORN IN THE LATE '60S**, and was still a toddler when the great protest songs of the "flower power" movement were in full bloom. It wasn't until years later that I came to understand the deeper meaning, and the motive power, of songs I'd grown up with. But when in history did popular music do more to change the world?

Take Crosby, Stills, Nash and Young's hit "Ohio." That song centered on the 1970 events at Kent State, when simmering tensions between Vietnam War protesters and the forces of law and order spilled over, and soldiers of the Ohio National Guard, responding to a campus protest, fired on a crowd of students, injuring nine and killing four. A photographer snapped a stunning, iconic image of a teenage protester on her knees by one of the bodies, open-mouthed in shock, and beyond distraught. The horror of that photo would go on to inspire Neil Young to write the powerful protest song "Ohio," focused around that very image. And the world sat up and took notice.

The soundtrack of the Vietnam War protest movement is legendary. Songs like Marvin Gaye's "What's Going On," Country Joe and the Fish's "I-Feel-Like-I'm-Fixin'-to-Die Rag," and Edwin Starr's "War," inspired by growing feelings of disenchantment, weren't grievances: They were calls to action. They worked their way into America's consciousness, changed minds, and ultimately helped bring a close to that protracted war. And over the years, other songs addressing other crises, from Peter Gabriel's "Biko" (about the death of an anti-apartheid activist in police custody) to H.E.R.'s anthem "I Can't Breathe" (about George Floyd's murder), have similarly sung truth to power, inspiring new generations by engaging listeners' emotions directly.

"I'd read about human rights issues and it didn't really seem to affect my life very much," says Peter Gabriel. "And then, when I was actually out there meeting people I started realizing that what I did had an impact, and could make a difference to things, encouraging everyone else to think and act."

Music is a powerful motivating force because it makes us *feel*. An angry protest song doesn't just inspire us, it empties us out of our chairs and into the streets. The roar of pump-up music at a boxing match or football game gets the crowd on its feet and shouting. An upbeat drinking song at the pub helps you bond with your mates. A fondly remembered holiday song brings you peace and nostalgia, and possibly induces you to call your mom. A stirring love anthem makes you turn that car around, right in the middle of the road, and race back to the airport to convince the girl not to get on that plane. Okay, that last one might be from an episode of *Friends*, but you get the idea.

Emotions aren't some mysterious poetic force—they're physical responses that have real purpose in your life. Think of them as a kind of evolutionary shorthand, designed to be triggered quickly and unambiguously by outside influences, like an attacking bear or a sexy stranger. Emotions help us shift quickly into various useful states—aggression, fear, bravery, intimacy—that help us fend off attacks, survive, and procreate.

And music's ability to not just express the songwriter's emotion (see Chapter 7: Create) but induce that emotion in others is precisely what makes it powerful... even dangerous. You don't get off your couch and join the protest because Neil Young is angry, you do it because *you're* angry—because his music magically transferred his feelings directly into you.

Music is the language of emotion; it is a highly evolved form of communication every bit as rich as the verbal one we use more consciously. From battle cries and alumni fight songs to elevator music and movie soundtracks, in this chapter we'll explore what feelings and emotions are, how music catalyzes emotional reactions, how easily emotional response can be manipulated (by the media, advertisers, retailers, and even the people we love), and how you can shore up your defenses and master your own emotions.

But before we start, we have to dive a little deeper into the basic concept behind all this: Why does music make us feel anything at all?

## YOU CAN FEEL IT ALL OVER

Emotion describes a shift in sentiment; the word itself comes from the Latin verb *emovere*, meaning "to move, or agitate." When we describe heartfelt wedding speeches and beautiful pieces of music as "moving," this is what we mean. A part of your mind that used to be over here...is now in a different place.

Some psychologists draw a distinction between emotions and feelings. Emotions, they say, are your body's automatic, default responses to an external or internal trigger. The instinctual reaction you have when you see a cute puppy, or discover a raccoon in your garage, is a chemical dance that's outside your control. Emotions

"I'm on top of the world, 'ey, waiting on this for a while now, paying my dues to the dirt."

**IMAGINE DRAGONS** "On Top of the World"

are automatic and brief; they come with unintentional facial expressions. Feelings, on the other hand, are your brain's interpretation of that chemical experience you just had. They come a little later, they're subjective ("I guess I feel more hurt than angry..."), and they can persist or mutate over time as your mind continues to be moved.

**FUN FACT**

Queen's "Don't Stop Me Now" was rated the world's most uplifting song by a study calculating scale, beats per minute, and other factors. All other songs bit the dust.

Neuroscientist Dr. Richie Davidson, a psychology professor and the founder and director of the Center for Healthy Minds at the University of Wisconsin–Madison, told me that what's most important to understand is that emotional responses to stimuli come directly from the limbic system, unfiltered by the conceptual and linguistic machinery of our frontal cortex. The limbic system is an ancient, instinctive, reactive, animalistic part of our brain, an alternative system for understanding the world that predates the development of spoken and written language. It gives us gut reactions rather than reasoned opinions, and we tap into it directly every day, though we're not always conscious of it.

Davidson invited me to consider smell as an analogy—a sensory input that similarly bypasses the controlling and categorizing mechanisms of our frontal cortex. When you encounter a strong odor, it goes straight to the limbic system for processing. You instantly understand the smell, pleasant or unpleasant, in exquisite detail. But you may have trouble describing it to others. "It smells a little like... it sort of reminds me of..." You'll struggle, as your forebrain tries to find linguistic analogies and context for this unique experience.

Music also directly engages the limbic system and produces emotional responses that aren't mediated or filtered by our thoughtful frontal cortex. This is why music can give you chills, make tears well up in your eyes, give you goosebumps, or make you sick to your stomach—these are physiological effects that have nothing to do with your brain's decision making. (Chills and goosebumps specifically, sometimes called **frisson**, apparently only happen to about half of us, those with strong connections between our auditory cortex and emotional processing centers.)

According to Davidson, music can affect our bodies directly like this *because* it bypasses the brain's usual filtering. "Music has those effects on emotions primarily because it's nonconceptual; it affects our mind and our brain at a level that doesn't involve our language," he told

me. "Our concepts are largely formed by language, and when we are immersed in our conceptual mind, it actually has the impact of disconnecting us from our emotions." In other words, music can bypass the organizational principles, like language and concepts, that we normally use to categorize new experiences and instead allow us to directly experience the emotions of others, just as you directly experience a smell. Says Davidson: "I think that many people are drawn to music because it provides a brief opportunity to go beyond the conceptual—to be, if you will, *relieved* of the conceptual."

So: The immediate emotional impact of a song is provided by your instinctual, reactive limbic system. As far as how you *feel* about it—do you like this song, are the lyrics meaningful to you, and so on—that's where your conceptual brain takes over. And here, the interesting thing to note is that the song doesn't impose its structure on the brain; the brain imposes structure on the song. Here's how:

Remember Rorschach tests, where a therapist would show a series of images made by random, swirling ink blots and invite the patient to tell her what he sees?

At first, it looks like a messy swirl of black and white smears, which is no surprise: That's what it actually is. But gradually an impression appears: The patient decides that this is a pumpkin. Or a water buffalo. Or two angry parents shouting at each other.

Recall that your brain is always hungry for patterns and meaning, and that when it doesn't find any, it invents some. There's a well-known phenomenon called facial pareidolia, for example, in which people "see" faces in inanimate objects. Glance around the room right now and find the nearest electrical outlet—doesn't it look a little like two shocked faces? We see faces where there are none because faces are very important to humans; our helpful brains would rather risk a misunderstanding than fail to recognize a possible threat. And a similar effect may help explain how music activates emotion.

To begin with, musical notes and chords aren't inherently "happy" or "sad"; they're fundamentally neutral sounds, arranged in patterns. Many musicians argue that minor **keys** are tinged with sadness, and major keys are happy, an idea based on the impression that the harmonic overtones of major chords sound natural (= happy) whereas minor chords, which include tones not part of the natural harmonics of a basic note, sound "off" to us, inducing us to feel disrupted, disconnected, or sad.

But evidence is emerging that this commonly accepted idea is a Western-world phenomenon driven not by anything inherent to the sounds themselves but by cultural history. When researchers from Durham University played major and minor sequences to Khow and Kalash tribespeople in Pakistan, the sequences failed to induce those specific emotional reactions. Conversely, tunes the tribe considered solemn sounded downright cheerful to Western ears. If minor-chord progressions sound sad to you, it may be because your musicians have been writing sad songs using those chords for a long time, and that's built up an expectation in you—your brain is imposing something on the song that isn't there.

"Major and minor, happy, sad is culturally specific to us in the West," says Levitin. "If you listen to Chinese opera and you've never listened to it before, or to music of the Cameroon Pygmies, you might be moved by it, but you'd be hard-pressed to say what emotion it's trying to convey, because that's culturally embedded."

Another historical example of this is the strange history of the tritone, an unusually high, three-tone step in music that you can hear for yourself in the first two syllables of "Maria" from *West Side Story*, or when you hear "The Simp..." in the theme song to *The Simpsons*. Today, a well-placed tritone is freely used to add a touch of unresolved strangeness to a song. But in the Renaissance, musicians across Europe avoided the tritone like the plague—and they knew a thing or two about the plague. They believed the tritone evoked feelings of fear and disgust; the sound was considered so dissonant and ugly that some suggested it invoked Satan himself. By the 19th century, feelings around "the devil's interval" had shifted. The dreaded interval never changed, but the cultural response to it did.

"The power of music to suggest, evoke, or induce particular emotions is culturally determined," Levitin explains. "It's based on a lifetime of listening to music, of soundtracks. If you hadn't really listened to music and you didn't know what the cultural conventions were, it wouldn't make any sense. If you know nothing about architecture, and an architect points you to a building and says 'Look at the way they've combined the rococo with the modern, and I see elements of I. M. Pei...,' you're going to say 'What? It's just an ugly building.'"

The point is: There is nothing inherent in the notes themselves that creates the rich emotional associations music has for most of us. This is not to say music doesn't evoke real emotion—of course it does. But it's culture and history that make the connection. Even at the gut level of a limbic system response, our ability to make sense of music depends on our experience. Music can hijack your emotional response system, but it isn't simple; playing happy songs doesn't always make you happy. For a more nuanced understanding, we have to look at musical communication.

## WHILE MY GUITAR GENTLY WEEPS

Anyone who has loved a pet or raised an infant knows that emotions can be communicated directly, even when speech is impossible. Long before they can speak, babies can read and express things like love and anger and trust—they "speak" emotion before words are an option.

# "Careful what you take for granted

# 'Cause with me, know you could do damage."

H.E.R. "Damage"

"Particularly early in life, when there is no language, the language is the language of the emotions," says Richie Davidson. "And of all the sensory channels available, sound is probably the most powerful vehicle for communicating emotion."

As we touched on in Chapter 5: Connect, music is thought by some to have developed as an alternate communication channel that predated the development of speech. We hum to our babies to tell them they're safe, we put them to sleep with lullabies, and, as we begin to coax them into the world of language, we show that words are communication by using a singsongy voice: "The wheels on the bus go round and round..." Ultimately those babies learn to speak, and verbal language becomes our dominant mode of communication. But our ability to communicate emotions through music never disappears.

In fact, music is built right into our spoken language. "Anthropologists say that about 90% of your communication is actually not the words, but the music of the words," says Helen Fisher, senior research fellow at the Kinsey Institute and chief science advisor to Match.com. She spoke passionately about the power of *prosody*, the way we enrich our spoken language with extra meaning through inflection, rhythm, pausing, and other microtechniques. Think about how you'd say the phrase "I love you" in these scenarios: a) to your significant other for the first time, b) to your mom on the phone, and c) sarcastically to a friend who just insulted you. Prosody lets us bring music to verbal language—so we can convey a lot of important emotional information that's hard for words alone to carry.

Fisher told me of a theory (not hers) that animals typically share three basic communication types that can be characterized as the whine, the bark, and the growl. A growl is obvious—it's a threat or a warning. A whine is like the sound an infant makes; it's passive, designed to neutralize a threat or confuse a potential attacker. And barking is the rest of our chattering transfer of information (picture the "conversation" in a busy dog park). It's an interesting construct that hints at how words might have evolved from preverbal kinds of communication like music.

For some, toward the end of life, when speech fails, they return to music for communication. "Music can lift us out of depression or move us to tears—it is a remedy, a tonic, orange juice for the ear," said the late, great author, neurologist, and professor Oliver Sacks. "But for many of my neurological patients, music is even more—it can provide access, even when no medication can, to movement, to speech, to life. For them, music is not a luxury, but a necessity."

Written and spoken language is humankind's great differentiator; it's made the modern world possible. Nevertheless, neuroscientists point out that language-first thinking can constrain our understanding of the world, by reflexively comparing each new experience to words and concepts we're already familiar with. And that's where music can play a meaningful role. "We often are drawn to activities that at least temporarily can remove us from the shackles that language can impose," says Davidson. "Music is a medium through which we can bypass the conceptual mind and activate emotion more directly."

So if music is the language of emotion, what are the rules of that language? When we're children, our pliable young brains absorb more and more music, and start to build a scaffolding of neural structures to help us identify similar musical patterns in the future. In the same

Whitney Houston's "I Will Always Love You" was Saddam Hussein's campaign song for Iraq's 2002 election. Guess what? He won.

way that we learn linguistic patterns that are "correct" or "incorrect," our brain builds a musical syntax that is specific to our culture: a set of implicit rules that, when followed, produce sounds that just feel *right*. As we intuitively learn the rules, our emotions become invested in these structures.

These structures, which comprise a pragmatic organization of our prior musical knowledge, impose their rules on the music we hear. We learn to predict where a song will go: a subconscious guessing game contributing to our satisfaction (when we guess correctly, as in a repeating chorus) and our excitement (when we are surprised, as in an unusual bridge). "Skilled composers manipulate the emotion within a song by knowing what their audience's expectations are, and controlling when those expectations will (and will not) be met," says clinical psychologist Malini Mohana on PsychCentral.com.

A reinforced expectation triggers our reward center; an unmet expectation may instead produce a slight sense of fear or danger and a hit of adrenaline. Soft, mellow music may match an internal expectation of a night winding down, or a childhood lullaby; your brain responds by modulating brainwaves down into a more restful state. Music with pounding bass notes matches an expectation that action is called for—here comes the adrenaline, and an associated feeling of power. And so on.

## THEY GOT ME HYPNOTIZED

The fact that music can alter your emotional state presents an opportunity for songwriters to make a connection. But it also opens the door for others to manipulate your emotions through music, with or without you knowing it. The same low bpm music that helps

# Why On-Hold Music Is Crappy

The first elevators, invented in the early 20th century, were absurdly scary things. ("Okay, here's the concept: We're going to shut you into a steel closet with no doorknobs, and use a thin cable to drag you hundreds of feet up inside a long vertical shaft high above the city streets. Who's first?") And for almost as long as there have been elevators there has been "elevator music": light, upbeat, typically lyric-free pablum. Why? Because that sort of calming music keeps boredom at bay and soothes the jangled nerves of elevators' not-yet-dashed-to-pieces inhabitants.

The same concept evolved into the "hold music" played while you're being endlessly rerouted through automated call centers: It's designed to confine your rage to a low simmer. As the story goes, a factory owner named Alfred Levy was taking a call in the 1960s when an exposed wire nearby caused his receiver to pick up a local radio broadcast. It also made a lightbulb go off in his head: Levy soon submitted a patent for hold music, believing a stream of light, upbeat, lyric-free music would keep callers on the line. Hold music is irritating, but silence is worse: It makes callers feel abandoned. A 2009 study showed that hold music makes wait times feel shorter. Your call is very important to us, so please be patient; the next available agent will be right with you.

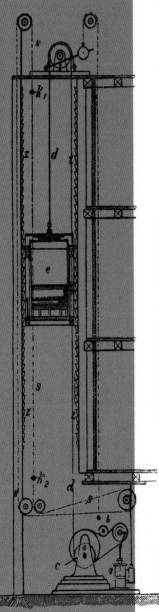

**FOUR QUESTIONS FOR**

# HANS ZIMMER

Widely regarded as one of the world's most innovative musical talents, Hans is a visionary theatrical composer, having scored some of the world's most successful films, including *The Lion King*, *Interstellar*, and *The Dark Knight*.

**Do you have a favorite genre of film?**

My favorite genres are the movies that actually nobody wants to go and see, which are movies about human crises, politics, Amnesty International, colonialism—all those unsavory subjects. So what I do is, of course, I cheat. I try to find directors I work with who have a similar sort of humanistic soul and heart beating inside them. And we disguise our little messages. We try to add substance and humanity to things that might have been lacking at the time when we first got to it.

What are the films that I love the most? The films that ask questions.

**Is your music an answer to some of those questions?**

No, I think it's a further way of asking questions, further ways of disturbing the status quo, further ways of finding out that maybe what you see on the screen is not necessarily the answer to everything or what people are saying. And certainly what you're feeling right now.

The music doesn't manipulate in a way whereby I make people feel something that they aren't naturally feeling, as opposed to just opening the door and going, "Here, I'm giving you an opportunity to feel something." I think that's actually quite interesting in the modern world, to take pause, to just feel—something. I think that's unusual nowadays. In between texting and answering your cell phone and taking a selfie, they actually get to feel something.

**How important is it to you when you get involved in scoring a film?**

Usually what happens with me is I get involved very often in this sort of unfinished draft, not a first draft. For instance, *Interstellar* was very much director Chris Nolan coming to me at one point and going, "If I gave you one page, but didn't tell you what the movie was about, would you give me one day and write whatever came to you on that one page?" I wrote this very fragile piece, because he had written something beautiful about what it meant to be a father. He came to listen to it at 10 o'clock at night on a Sunday. And I said, "Well, what do you think?" And he goes, "Well, I better make the movie." And I went, "What's the movie?" He described *Interstellar* and all its epicness and its glory. And I said, "But all I wrote was this tiny fragile piece." And he goes, "Yes, but I know where the heart is." So sometimes it's all about the first draft or whatever, it's a constant conversation.

**Do you have an example of a go-to sound to evoke a powerful moment?**

It gets interesting when you do it the other way around. One of the first scores I ever did in Hollywood was *Driving Miss Daisy*, which ostensibly is really quite a conservative film about an elderly lady. And if you really listen to it, it's an electronic soundtrack, with the occasional banjo thrown in.

It's nice having cultures collide, it's really interesting. With music, we can do it all the time where we gather musicians from many different tribes of the world, and don't just stay within the confines of the European Western orchestra. The great thing about film music is there are no rules. You can go and do your psychedelic-country-western-electro-punk track and nobody's really gonna question that. They might not like it, but it's perfectly legitimate to do that.

And I was told by everybody that everybody's attention span is zero. You know, don't do anything over three minutes long or you're dead. But the weird thing was that film music is long because it tells a story. And it goes through all these mood changes, et cetera. I mean, when I do *The Dark Knight*, it's the 22-minute piece. Where I do *Pirates of the Caribbean*, it's the 14-minute piece.

They're long pieces and people love to be put into this world, and they have the patience for it. And, in a way, we must stop listening to those people who tell us that we can't do it.

I think if we can make the world a little quieter now, and curate the world a little bit, we can reach people in a whisper, in a delicious, beautiful, intimate, heartfelt whisper.

us relax also makes us more suggestible. For example, in order to get her patients to open up, a psychiatrist will have them recline on a comfortable couch in a quiet private room to create a safe space and subconsciously lower their fear. In a similar way, relaxing music can put you in a state where you're more compliant and agreeable, whether you want to be or not—a fact that big business, particularly in the age of big data, has lost no time learning to exploit.

"Fast-paced music makes you buy more things," says Heather Berlin. "They're trying to get you to be out of yourself and not thinking so much...if you get in a rhythm and you kind of dissociate a little bit, you can make decisions that are more automatic, without rational or contemplative thought."

To be clear, this isn't theoretical; this is literally happening all the time to you, right now. Here are just a few completely true examples from this week (your experience may vary):

- Your wine store plays classical music, because it makes people buy more expensive wine than when it plays Top 40 music.

- Grocery stores and restaurants keep the tempo light and slow, because it makes you linger longer and order more.

- Bars play loud music to induce you to talk less and drink more.

- Coffee shops speed their music up at the end of the night to subconsciously motivate you to go home.

- Call centers and elevators play light, lyric-free music to keep you calm. (*See sidebar, page 189.*)

The effects are even less subtle on TV and in the movies, where a professional class of people called producers consciously try to craft a comprehensive emotional experience for audiences, with music no small part of it. "The media to which we're exposed often has a

"I think the moment chooses the anthem and not the anthem choosing the moment."

QUESTLOVE *on what makes a song historical*

soundtrack," notes Davidson. "And that soundtrack likely plays an extremely important role in the impact that the media has. I think if we were exposed to the identical information without any sound, in the media, this world would be a very different place."

# Finland has the most heavy metal bands per capita in the world. Take that, Turks and Caicos!

For any activity in which you're involved, public or private, if there's music in the background, you have every reason to be suspicious that you're being manipulated.

Music's role as subliminal mood manager has a long history. It was a World War I major general, George O. Squier, who first hit upon the idea that background music could be piped into work settings to enhance productivity. As the day progresses, he mused, workers get increasingly tired—but spurts of increasingly buoyant background music might keep them motivated. To bring this idea into practice, Squier's company introduced a playlist, called the Stimulus Progression, that played increasingly peppier music as the day went on. In the 1950s and '60s, this "whistle while you work" music became the soundtrack of capitalist America.

The fascination would continue for decades. In the early '70s, researchers discovered that TV commercials with carefully chosen background music made people more likely to rate the advertisement more highly. In 1986, a team of researchers analyzed a thousand commercials and found that music functioned as an "auditory memory device," improving brand awareness. One study found that "happy music produced happier moods in subjects, but sad music produced the highest purchase intentions." Another discovered that, in supermarkets and restaurants, people lingered more and bought more when the tempo of music was slowed to around 72 bpm.

Over in Hollywood, music's power to move audiences was not news; the movies had already been using soundtracks to modulate audience emotions for decades. In 1933, audiences of *King Kong* heard the very first movie score, by Max Steiner, designed specifically to support the story at every turn. There was a monstrous Kong motif (brooding, dangerous, masculine), pre-terrifying audiences with the promise of doom and destruction, and a heroine Ann motif (described as "lyrical and feminine") for the damsel in distress; later, the two motifs would morph together into the love theme. The film was a smash, and soundtracks and movies would never again be parted, from the dark, haunting themes of *The Godfather* to the mysterious, ringing celesta chamber music of *Harry Potter* that helped audiences feel deeply immersed in a world of enchantment.

After *King Kong* it took only two years for Best Scoring to be added as an Academy Award category. Today, the award show's most-nominated living person is a soundtrack composer, John Williams, with 52 nominations, including such small, forgettable art-house films as *E.T. the Extraterrestrial, Star Wars, Home Alone, Raiders of the Lost Ark,* and *Harry Potter and the Sorcerer's Stone.*

Movie scores help us disappear fully into the story at hand, engaging our limbic systems so

that we literally feel what the characters feel. It's physically unsettling, watching *Jaws*, every time you hear those famous shark-is-coming tuba notes. We feel terrified at the shrieking violins ("Eeek! Eeek! Eeek!") during the stabbing in the shower in *Psycho*, elated as Rocky runs up the steps of the Philadelphia Museum of Art, hopeless as the ship goes down in *Titanic*. Music soundtracks manipulate the hell out of us, but they give us the ride of our lives.

"The juxtaposition of music with a scene can create ambiguity resolutions," says Levitin. "I

think the first time this was used in a commercial film was in 1967's *Bonnie and Clyde*, when Faye Dunaway, Warren Beatty, and Michael J. Pollard are shooting up a bunch of people and it's a bloody mess. This was probably the bloodiest feature film ever up to that point, and they play this sort of jaunty music...and what that does is it tells the viewer this is how Bonnie and Clyde are seeing it."

Thanks to carefully applied music, you feel the same emotions that these out-of-control gangsters feel. And at the end of the day, *that's* what disturbs you.

# CONCLUSION

Music is the language of emotion, with a nonverbal impact that bypasses our clever **cerebral cortex** and communicates directly with our ancient, emotional limbic system. We feel what the composer or songwriter wants us to feel before we understand what she's trying to say. "Music can impact us profoundly, shaping or changing our moods if we let it, and so we have to be in control of what we choose to listen to and use music instrumentally for our purposes—not the other way around," warns Francesca Dillman Carpentier, media psychologist and professor at the University of North Carolina at Chapel Hill.

Music is your kryptonite; its ability to play with your emotions provides a convenient backdoor for unscrupulous store owners and dating-app Lotharios to hack into your brain's emotional center and manipulate you. But you can fight back. The first step, naturally, is awareness: Understand how and why music is being used to influence emotional decisions— and maybe consider wearing influence-canceling earbuds now and then.

You can also make more active use of music to regulate your own mood. Consider working actively on the soundtrack of your life, weaving carefully chosen music into the fabric of your world. From empowering alarm clock wake-up songs to kick off an important workday to restful dream generators as you slip back into bed, you can leverage music's magical power to jump-start the moods most conducive to perfecting your day and to better arm you to resist outside influence. The best defense is a good offense. ■

# TAKEAWAYS

*Music is a potential weak point in the defenses of our otherwise well-protected brain—the uncovered exhaust vent to our Death Star, if you will—and it's certainly used all the time to motivate and manipulate us. It can turn us into eager consumers, obedient workers, shaken and stirred moviegoers, or energized soldiers. But if we are mindful of how easily music can affect our emotions and decisions, we can go about our lives with clearer heads and use music's power over emotions for good.*

### TO MAKE PEOPLE LINGER
At your party, your store, or your restaurant, opt for music in a slow-ish but not sleep-inducing range of around 72 bpm (like "When I Was Your Man" by Bruno Mars). Sounds trite, but slow music really does slow people down.

### TO GET PEOPLE TO GO
Conversely, if you want to motivate them to move more quickly or get their butts home, speed up the playlist with songs whose frequencies are at least 85 bpm (like "The Longest Time" by Billy Joel).

### TO COMPOSE SONGS THAT PACK AN EMOTIONAL PUNCH
Toy with your audience's musical expectations. Our emotional response to music is rooted in the brain's ability to predict where the tune is headed; generally, enjoyment comes out of the right mix of meeting and surprising those expectations.

### TO SHAKE A BAD MOOD
Carefully build and execute a playlist that starts sad (acknowledging your mood) but gradually shifts to upbeat, familiar songs. Your body's ingrained emotional responses will lead your conscious mind and help change your mood.

# BECOME

# DE DO DO DO
# THAT'S ALL

**THE POLICE**
"De Do Do Do,
 De Da Da Da"

# DE DA DA DA DA
# I WANT TO
# SAY TO YOU

**TRIANGLE ENTHUSIAST PYTHAGORAS** thought a lot about music. He studied the lyre, a stringed instrument that was kind of a handheld harp, and developed the mathematics around how tone changes with string length. Pluck on a lyre string and you contribute energy to it, making the string vibrate at a particular frequency that's a function of its length; your ears experience that vibrating energy as a specific tone, like B-flat. Pythagoras believed that everything in the natural world was similarly ruled by math. When he looked up at the night sky, he didn't just see stars and planets—he saw an elegant equation at work. He believed that each planet emitted a different sound as it rotated and revolved, like the strings of a lyre, each contributing a unique celestial tone to a beautiful cosmic symphony. This concept became known as "the music of the spheres."

There is a different sort of string theory under consideration today in the realm of theoretical physics. When we sliced open molecules, we found atoms; inside atoms we found protons, neutrons, and electrons;

and inside protons and neutrons we found quarks, which seem to be indivisible. String theory surmises that if you could zoom in all the way on a quark—which you can't, even in theory, because they're smaller than the wavelength of visible light—what you'd find is not a set of even tinier particles but rather an infinitesimal vibrating string of energy. And it's the clustering of these micro-vibrations—these tiny, self-strumming lyre strings—that gives rise to the illusion of solid matter. This energy is everything there is: the endless dance of Shiva in Hindu mythology that gives substance to the world. This is the music of the universe, and your body and your mind are a part of it, too.

I hate to make a lyre out of you.

It sounds poetic, and maybe even a little bit corny, but music is in you. It's built into the fabric of your atoms, and etched deeply into the structures of your mind. Music has spent tens of thousands of years slowly modifying your DNA to build a brain and body more receptive to its influence. And over eight chapters detailing the fortuitous intersection of music and mind, we've tried to deliver insights on how to leverage music to sleep, focus, love, thrive, connect, escape, create, and feel. But the back cover is coming up quickly now, and it's time to decide what you're going to do with all this knowledge.

In this final chapter we'll explore music's power to improve not just how you live but who you are. We'll

> "We have no past, we won't reach back, Keep with me forward all through the night."

**CYNDI LAUPER** "All Through the Night"

look at music's central role in expressing your identity and personality, including elements of spirituality and faith. We'll show how music can help you make lasting changes to behavioral patterns, from fixing bad habits to building stability and happiness. And we'll explore how you can leverage music to deliver inner peace and find your purpose, helping you become...well, whatever the hell you want to be.

Music is deeply entertaining, and for many people that's enough. Most of the hairless apes milling around you will enjoy music throughout their lives without ever realizing its potential to change their minds and change their fortunes. But if you're ready to step out of the comfort of the Pandora algorithm and into the deejay booth, there's a world of opportunity at your fingertips.

## WHO ARE YOU?

"Music is one of those activities that seems to combine more of our human facilities than anything else," says Tod Machover. It soundtracks your weddings and funerals, yes, but also your commutes and happy hours; your promising make-out sessions, tearful breakups, and joyful reunions; your elevator rides and drives to the dump. It's the cosmic background radiation of our daily lives. Or, as Youssou N'Dour puts it: "When I hear someone speaking something, my mind translates that voice into music. My entire environment and everything around me—people talking and people praying in the morning—become music for me. Everything I hear, I hear as music. It's my way of life. If there is no music, I don't feel free."

Like any constant companion, music enjoys an outsize influence over your personality. All along the journey, your music is quietly shaping you: firing you up and calming you down, cementing some memories at the expense of others, cuing emotional responses, resurrecting your past. "You can't get away from the fact that everything has a rhythm to it in its most simple form," says Mick Fleetwood. "It's that powerful. It's like sitting in front of a wave." Your mind hungers for rhythmic inputs, and when it receives them, it reprograms itself around them. You *are* your music, in the end.

And as a result of this influence, the music you love telegraphs deep truths to the world about who you really are. In fact, research has shown that total strangers can make accurate detailed judgments about you—your open-mindedness, your creativity, where you are on the introvert–extrovert spectrum—after listening to just 10 of your favorite songs.

In a pair of groundbreaking studies, curious University of Cambridge researchers paired people up in a room and gave them a simple task: Get to know each other. They could talk about anything they chose, but the most popular topic was music. When study participants needed a conversational shortcut to a stranger's personality, their favorite opening gambit was musical preferences. Why? Possibly because they knew instinctively that it would work—because our musical tastes say just about everything worth knowing about us.

Take empathy, for example—a prized trait highly worth discerning in a stranger. Humans tend to cluster around one of two cognitive styles: They're either empathizers (more likely to make decisions that take others' feelings into account),

Music could soon help firefighters. Turns out sound waves between 30 and 60 Hz can separate flame molecules from their surrounding oxygen, putting out the fire.

complexity, because they're better prepared to appreciate it. By and large, we clever humans are pretty good at constantly assessing the people around us for threats and opportunities—this is just a hint of the subconscious mental math behind some of those thought processes. "Maroon 5 concert T-shirt? He'd probably be a good listener. On the other hand, that leatherhead cranking the Visigoth tunes might be just the jolt of crazy I need right now..."

Many of us are instinctively comfortable using someone's musical preferences as a starting point for getting to know them. Quick experiment: Imagine a new hire, Pat, is joining your small company next week, and you're curious whether you'll approve. Your begrudgingly sympathetic boss has agreed to tell you just one thing about the new person. You can get the answer to one of the following:

- Pat's favorite binge is either *Ozark* or *The Masked Singer.*

- Pat's favorite reading material is either classic literature or graphic novels.

- Pat's favorite music is either electronic dance music or country and western.

Which feels like the most meaningful metric?

"Music is part of who you are," says neuroscientist Heather Berlin. "And you can use music to tap into these deeper parts of yourself that are beyond language, that

or systemizers (prone to basing their decisions on objective systems and rules). When researchers divided participants by cognitive style and looked for correlations with their musical tastes, they learned that people who tested highly for empathy also generally preferred mellow, relaxing music—soft rock, soul, indie, R&B—while those with systematic brains tended to prefer complex or intense music, like jazz or hard rock.

The point isn't that this is a groundbreaking insight—quite the opposite. You might very well have guessed that the high-scoring empathizers would gravitate toward music with emotional and lyrical

# WHAT YOUR MUSIC SAYS ABOUT YOU

*Researchers at Heriot-Watt University in Edinburgh surveyed 36,000 music listeners, looking for correlations between their personalities and their favorite musical styles. Here's what they found, as summarized in a Psych Central review of the data:*

| | Self-Confident | Creative | Outgoing | Introverted | Gentle | At Ease | Hard-Working |
|---|---|---|---|---|---|---|---|
| **BLUES & SOUL FANS** | ✓ | ✓ | ✓ | ✗ | ✓ | ✓ | ✗ |
| **POP FANS** | ✓ | ✗ | ✗ | ✗ | ✗ | ✗ | ✗ |
| **COUNTRY & WESTERN FANS** | ✗ | ✗ | ✓ | ✗ | ✗ | ✗ | ✓ |
| **RAP FANS** | ✓ | ✗ | ✓ | ✗ | ✗ | ✗ | ✗ |
| **HEAVY METAL FANS** | ✗ | ✓ | ✗ | ✗ | ✓ | ✗ | ✗ |
| **OPERA FANS** | ✓ | ✓ | ✗ | ✗ | ✓ | ✗ | ✗ |
| **CLASSICAL FANS** | ✓ | ✗ | ✗ | ✓ | ✗ | ✓ | ✗ |

you just can't get to in other ways." Songs operate on a visceral level, communicating emotions, intent, and values instead of words and ideas. They stir you, thrill you, crush you, or inspire you at a deep part of your mind where spoken language can't penetrate. That's why you can't always understand—never mind put into words—why *this* is your all-time favorite power ballad, or why *that* particular guilty-pleasure song (you know the one) secretly makes you cry.

Aside from inferring people's values from their musical preferences, we also reflexively use them to preemptively decide whether a stranger is "one of us" or not. Think back to high school. (I know, I know...this'll only take a minute, I promise. Or just skip this paragraph.) Whatever your school's social cliques were—freaks and geeks, goths and stoners, populars and brains, vampires and werewolves—they likely had specific musical genres they favored as a group. When someone blares "their" music on the subway or from a passing car, they're trying to tell you something, and it has nothing to do with their love for that particular song.

Says Dr. Jonathan Berger, professor at Stanford University, "There are commonalities in experience within cultural groups and subgroups that map aspects or attributes of 'music' to a particular group, while others (even within the same culture but, say, generationally separated) may call this 'noise.'" Music helps provide social groups with a tangible proof of commonality, simultaneously providing a beacon of identity—this is who we are!—and a wall to keep unbelievers out.

It's impossible to talk about group identity, and music's potential role in facilitating group bonding, without discussing oxytocin, the so-called cuddle hormone that makes you feel emotionally close to others. As discussed in Chapter 3: Love, oxytocin is a powerful potion for facilitating and strengthening bonds between people: young lovers, owners and their pets, parents and their children, college seniors and the Uber Eats delivery guy. And oxytocin can be released by music, especially music that's soothing and/or sung aloud. In fact, oxytocin is just part of a delicious cocktail of chemical reinforcement that also includes endorphins and dopamine, which combine to help make music a powerful glue for group cohesiveness.

The fact that shared music is a potent signifier of group identity may help explain why music is such a stubbornly universal badge of belonging: why countries have national anthems, high schools and rugby clubs have fight songs, armies have marching bands, and ethnic groups link arms and belt out drinking songs from the old country. But group identity isn't all unicorns and rainbows. The same music that bonds a family, community, or organization together does so, after all, by explicitly excluding others. Declaring an "us" necessarily defines everyone else as "not us," and the closer you are to your in-group, the farther you are from everybody else.

Which leads us, naturally, to religion.

## "If there is no music around, I don't feel free."

YOUSSOU N'DOUR *on music as a life force*

# LOSING MY RELIGION

The ability of music to move people has frequently made it the subject of controversy. When religious music started appearing in Christian church services, back in the fourth century CE, Augustine of Hippo (later Saint Augustine) protested the onslaught, arguing that the music brought on so much pleasure that it would distract worshippers from the divine. Churchgoers, he feared, would get so caught up in the music that they would forget to focus on the lyrics. Remember, this wasn't the devil's heavy metal Augustine was protesting—these were *hymns*, specifically created by religiously inclined people to honor God. But the power of music threatened to compete with the power of the Word.

For all his misgivings, Augustine actually loved music—in fact, it was its fearsome emotional power that scared him. He confessed to personal reactions like "The music surged in my ears, truth seeped into my heart, and my feelings of devotion overflowed." Augustine believed that, if used wisely, hymns sung by a group could bring a congregation closer to God, and eventually he overcame his objections and encouraged religious leaders to embrace music. And that's why there are thick, bound singalong hymnals on the back of every Christian pew on the planet.

Augustine's ideas would be extremely influential in the Catholic Church, specifically in the creation of a large canon of hymns. For the growing church, hymns became a fundamental way of expressing religious identity, uniting millions of followers in a global community, all singing along, across distance and over centuries. And music has played a similar binding role for other religions, and countries, and sports team fan bases, and countless other cultures and subcultures.

Religion has known benefits for the brain, but music changes the equation. When Catholic nuns pray and Buddhists meditate, their frontal lobes (the home of concentration) light up with activity,

# PETER GABRIEL

Peter Gabriel is a groundbreaking artist, vocalist, and humanitarian who powered Genesis, enhanced global consciousness of world music, expanded the boundaries of music video production, and has led human rights initiatives around the world.

**What is it about a protest song that gets people riled up?**

Well, most songs are about love me or f— me, then maybe they get beyond that to refer a little more to what's going on in the world. Then, maybe the best-intentioned end of it is "We could do something about the world and change it."

I guess the closest I've come to that is with the "Biko" song, which we did on the Amnesty tours, and it really was an emotional rallying point. At that time we were talking about apartheid, but in a way it became a theme tune for a human rights movement, as have other songs. I know quite a few people have written to me saying that after that concert... they decided to get involved in human rights and mention "Biko" quite often. So you only need one Mandela or Stephen Biko or Martin Luther King Jr. in an audience to maybe make a difference.

**What are your thoughts on the ability for music to bring humans and animals together?**

My own experience with bonobos totally changed my outlook on animal intelligence and their capacity to interact with us and understand our language, even though many scientists would still deny them that degree of intelligence.

And it was absolutely clear from my time with these bonobos, and with Koko the gorilla, there was a very high level of understanding of what was being said to them. And then I played music with them—it was just extraordinary and felt like I'd been let into some ancient history with my ancestors. This evolutionary gap just disappeared, and they wanted to communicate their feelings with me.

My dad was an engineer/inventor at a company that developed futuristic technology around television. They also had a Muzak division called Reditune. So he tried bringing some of this music into his milking parlor, because we had grown up on a dairy farm. So early on, the cows were given this calming music, and sure enough, it boosted milk output. There's no question that music can work to relax.

**What is it about rhythm in particular, especially through drumming, that has been so inspirational to you?**

I started playing music as a not-very-good drummer, so I've always loved rhythm. And for me, if I'm ever asked how kids should learn music, I always say start with the drums. If you get the grooves into your soul, everything else is easy.

When you put people on a group of drums and get them to play together, they start relaxing and smiling. And even if they're not very much in time, at a certain point, they just forget about being good at anything. They are part of it and join this big beating heart, which is the rhythm. It's a wonderful thing when you see it, when you speed it up above your heart rate, it takes you one way, and if you slow it down, it takes you another way and your body responds to that. It's a gear changer. But it's not just a physical beat changer, it's an emotional changer at the same time, because the different gears produce different feelings in you and in the listener.

**How does harmony affect a love song?**

Harmony, after rhythm, is the next tool you have to chisel your sculpture emotionally...Say a fifth interval, like a C and a G, will sound very nice together. You can look at the relationship mathematically; harmony has a lot to do with math. Say a major seventh interval, which you hear in warning signals in animals quite a lot, is the opposite. It wakes people up—it's an alarm. Every different harmony has a unique emotional response.

If you play a C and an F instead of a C and a G, it still sounds pretty good together, but it's not as sweet as the C and the G or the C and the E. You can carry on exploring these intervals and see how they make you feel. All good writers, good composers, know how to juxtapose these harmonies in a way that makes it sweet, interesting, stimulating, scary, terrifying, or just transcendental. I think harmony is a more sophisticated musical cocktail but still fundamental. If rhythm is foundation level one, then harmony is level two. Timbre is level three.

## What's the best meditation music?

*According to those who study psychoacoustics, the branch of science examining the perception of sound and its impact on experience, the most effective music to listen to while meditating might be binaural beats. If you recall from Chapter 1: Relax, binaural beats are basically a sound illusion in which two different frequencies are played, one in each ear, producing a song of a third frequency. It's deeply relaxing, and the effect can cause activity in different brain areas to synchronize—the perfect baseline for productive meditation.*

while their **parietal lobes** (responsible for sensation) quiet down. Prayer can cause people to relax deeply, even lose their sense of space or time. But Islamic prayer, which is much more musical, has a different effect. According to a report in the *Journal of Physiology*, Islamic worship reduces activity in the frontal lobes, boosting feelings of surrender and connectedness. "Islam means 'surrender' and its central idea is a surrendering to the will of God," according to Khan Academy; the sung prayers may help worshippers put themselves more reliably in the desired mindset.

History is rife with stories of music inspiring religious experiences all by itself. To explore how this happens, researchers for a 2018 study asked Muslims in Turkey to listen to Sufi reed music or German hymns played on the pipe organ. The reed music triggered deep emotions in many of the Muslim listeners, meeting the study's definition of religious experience. Curiously, though, the pipe organ—culturally resonant for Christians, but not for Muslims—had no such effect. The religious power of music depends, at least in part, on a person's cultural upbringing. As the study's authors put it: "Sufi music and religious experience was learned earlier by the individual and the same music serves as a memory prompt by reactivating religious experience and deep positive emotions."

Even for nonbelievers, music can deliver a kind of hyperpresent flow state very similar to that achieved by meditation. As discussed in Chapter 6: Escape, a song you love can enable you to step completely away from your daily worries and into an out-of-body experience. Here's how:

- A song, when you are focusing on it, requires your full attention across a span of time; it pulls you into its alternate reality, and therefore out of your daily grind.

- That song has a rhythm, awakening your brain to input, and induces emotions, to which your poor reactive brain responds as if the emotions were real for you—i.e., as if you were living the story in the song.

- And yet you *know* it's a song. There's no real fear, or pain, or crisis, so you get the joy of the ride with no consequences, and can happily soak up the dopamine in the aftermath.

Many of us busy cats indulge in our favorite music only in the context of doing other things, like driving or cooking—as background noise for our busy life. But what if, instead, you carved out a regular, meditative space for enjoying positive, affirming music that you personally enjoy? Like taking a regular daily or weekly yoga break, but with headphones...making a commitment to spending just half an hour or so to enjoy music's unique ability to help you forget the past and ignore the future, and just be completely immersed in the here and now.

If one such experience is a pleasant escape—and you know, from your own experience, that it is—imagine the

cumulative effect of indulging in this kind of musical meditation on a regular basis over a lifetime. There's evidence, through studies around a similar experience called loving-kindness meditation, to suggest that over time this kind of regular out-of-body experience can actually build you into a happier, more satisfied person. As quoted in the online journal *Quartz*, the study's authors said: "The practice... led to shifts in people's daily experiences of a wide range of positive emotions, including love, joy, gratitude, contentment, hope, pride, interest, amusement, and awe...They enabled people to become more satisfied with their lives."

Sounds like heaven to me.

## THIS IS GONNA BE THE BEST DAY OF MY LIFE

Even when we're not consciously paying attention to a song, we know that our subconscious mind is processing it anyway, sifting through the lyrics with the same seriousness as if the singer were speaking directly to us. The question is, just how susceptible are we to being

subconsciously influenced by music and lyrics? That's not to say that listening to happy music automatically makes you happy—sometimes it can make you want to put a brick through the window of a Chuck E. Cheese. But take a look at a couple of things individual studies have found to be true:

- Fast music encourages listeners to walk faster; slower music slows people down.

- Aggressive lyrics increase aggressive thoughts and feelings.

- Listening to happy music facilitates divergent, creative thinking.

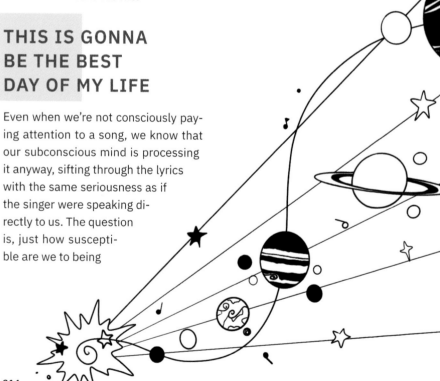

- Sad music further depresses adolescents who are already depressed.

- Hearing music you like decreases stress and makes you happier.

Are you seeing a pattern here? We are deeply affected by our music: not just in theory, not just during music-therapy sessions, but on an everyday basis. And because music is an ever-present factor in your life, it's changing your life all the time, whether or not you intend it to.

Remember the pivotal scene from *A Clockwork Orange* where the aversion therapy of the evil social scientists causes the main character to associate his favorite artist, Beethoven, with violence and disgust? Well, there's no easy way to put this, but we're getting closer to harnessing this terrible power. A team of researchers from two Canadian universities discovered that by deploying magnets in particular places near a subject's

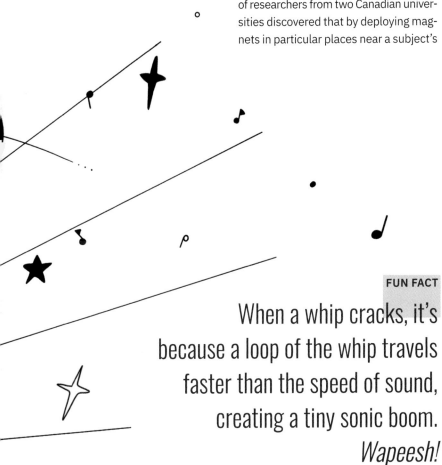

When a whip cracks, it's because a loop of the whip travels faster than the speed of sound, creating a tiny sonic boom. *Wapeesh!*

head, they could alter the strength of the reward response and actually *change a person's preferred genre of music*...at least temporarily. That is a recipe for some cruel and unusual punishment.

It also represents an incredible opportunity. If your musical preferences are themselves potentially malleable, then nothing's written in stone. For the first time in human history, you have tens of millions of songs literally at your fingertips, and a clearer understanding of how to leverage them to your advantage. Time to use this unprecedented toolkit to rewrite the story of you.

"We all have this narrative that we return to over and over again," Richie Davidson told me. "And for some people, the narrative is really all they have: It is their entire identity, and it's very difficult for them to step back and see it as a narrative; they just experience it as who they are. But once a person is able to step back and see this narrative for what it is, then the mood can subside and be more easily regulated."

Music can be the breakthrough you need...just press play.

In the end, it's all about **reverberation**. "Everything in the universe has a vibration and frequency," says Susan Magsamen. "So we are filled with vibrations all the time. And our ability to feel them, even if we may not be able to hear them, is amazing." The strings of your life are vibrating even now, down at the atomic level. Your music gets into your head, literally and metaphorically, and resets

## "I can tell you that every single thing that Philip Glass ever wrote can be ice-skated to."

LAURIE ANDERSON *on Philip Glass's true genius*

# The Curious Case of 40 Hz

*Medicine and surgery cannot currently reverse Alzheimer's disastrous effects. But sound and light may be able to, according to Tod Machover, a professor at the MIT Media Lab, thanks to gamma wave entrainment. The gamma frequency, highest of our five natural levels of brain activity, is associated with activating functions around mental sharpness, and stimulates what can be thought of as our brain's immune system. And these roles make it an extremely promising treatment for Alzheimer's.*

*Specifically, a frequency of 40 Hz (which corresponds roughly with the low-E on a piano) is known to provoke activity in microglial cells, which destroy pathogens in the brain and remove damaged cells. As with humans, the microglial cells of mice with Alzheimer's can't clean up the plaque debris that causes Alzheimer's—and their brains display little or no gamma activity. But by exposing the mice to gamma frequencies through light and sound, researchers were able to reactivate those microglial cells, remove plaque, and reverse some symptoms of Alzheimer's.*

*Will it work in people? Machover's team, following on from discoveries by Li-Hsuei Tsai and Ed Boyden, also of MIT, has seen incredibly positive early results. "Our work shows that humans can be successfully exposed to sound or light at 40 Hz—with especially surprising results when both are combined—in a way that kickstarts the brain into producing gamma frequencies again," he reports. Working in concert (literally and figuratively) with other MIT teams, Machover's "Opera of the Future" team is working out the best ways to build this "gamma therapy" into specific musical works that please the ear and can ameliorate mental issues including Alzheimer's, Parkinson's, and depression.*

the rhythm of your life in countless ways. On a daily basis, this means better sleep, tighter focus, and more efficient workouts. Over the long haul—with a little informed choreographing from you—music's effects can catalyze meaningful long-term improvements in your health, behaviors, sanity, relationships, fortunes, and personality, and literally help you become the person you want to be.

# Music has been called many things: history, the stuff of love, what feelings sound like. What's YOUR definition?

It's not just theoretical: Science, tech, and medicine are converging to help you fully leverage the power of music like never before. "Imagine taking a snapshot of your brain and immediately having a customized music library tailored to your specific neuroanatomy," says Diana Saville, discussing the burgeoning field of computational neuroaesthetics. "With playlists to help you brainstorm like never before, achieve a zen moment, inspire awe, work more efficiently, power down for a nap, get hyped for a party, make healthier choices, to be a social butterfly, and on and on. I'd sign up for that!"

The secret behind the unique and wonderful soundtrack of your life is that your music has been playing you all along. But with a little willpower, you can take control of the mix, and become your own driving force toward whatever and whoever you want to be.

The only question left is: What are you going to put on next?

# TAKEAWAYS

*Whoever or whatever you want to be, music can probably help you get there. Your playlist is a pretty good predictor of who you are, but that's only its passive state. Leveraged actively, it can also be an extremely powerful agent for changing who you are.*

**TO FORGE STRONGER SOCIAL BONDS** Try singing in a choir, or sing along during the hymns at church. Group singing has been shown to improve social engagement, and boosts well-being and overall health.

**TO BE MORE GENEROUS** Listen to music with uplifting messages. "Prosocial" music, with lyrics about making the world better (like John Lennon's "Imagine"), breaks down biases and fosters "helpful behavior," according to several studies.

**TO MAKE NEW FRIENDS** learn to dance. Synchronized movement combines with the releasing of endorphins to make us more cooperative and more receptive to new relationships.

**TO BECOME A MORE CREATIVE PERSON** Play happy tunes. Listening to songs designed to engender a positive mood also makes you more likely to engage in the divergent thinking and creative cognition needed to solve today's complex problems, according to researchers.

**TO BECOME MORE COMPASSIONATE** Learn an instrument and stick with it. Musicians score more highly than the average person for the complex cognitive and emotional trait of empathy, perhaps because making music strengthens connections all over your brain, including between your "thinking" forebrain and your "feeling" limbic system.

# SCIENTISTS

## INTERVIEWED FOR THIS BOOK

**JOY ALLEN**

Dr. Joy Allen is the chair of the Music Therapy Department at Berklee College of Music, the founding and acting director of the Berklee Music and Health Institute, and an expert on the intersection of music and health and wellness.

**JONATHAN BERGER**

Dr. Jonathan Berger is a professor at Stanford University, with research areas including music, science, and technology. Dr. Berger specializes in composition, music theory, and the exploration of cognition, consciousness, and conscience.

**HEATHER BERLIN**

Dr. Heather Berlin is a neuroscientist, clinical psychologist, and associate clinical professor of psychiatry at the Icahn School of Medicine at Mount Sinai. She is also a host on the *StarTalk All-Stars* podcast with Neil deGrasse Tyson.

**FRANCESCA DILLMAN CARPENTIER**

Dr. Francesca Dillman Carpentier is a media psychologist at the University of North Carolina at Chapel Hill, conducting research on the effects of mass media, audience responses to depictions of sex in media, and audience motivations governing media content choices.

**RICHIE DAVIDSON**

Dr. Richie Davidson is founder and director of the Center for Healthy Minds at the University of Wisconsin–Madison, as well as the founder of Healthy Minds Innovations,

a nonprofit associated with the Center. He is best known for his groundbreaking work studying emotion and the brain. A friend and confidante of the Dalai Lama, he is a highly sought-after expert and speaker, leading conversations on well-being on international stages such as the World Economic Forum, where he serves on the Global Council on Mental Health.

**LAURA FERRERI**

Dr. Laura Ferreri is a professor at Université Lumière Lyon 2. Her research areas include music and cognition, music and dopamine reward response, and memory.

**HELEN FISHER**

Dr. Helen Fisher is an anthropologist and expert on romantic relationships. She is a senior research fellow at the Kinsey Institute for Research in Sex, Gender, and Reproduction, and is the chief science advisor to internet dating site Match.com.

**ADAM GAZZALEY**

Dr. Adam Gazzaley is the David Dolby Distinguished Professor of Neurology, Physiology and Psychiatry at University of California San Francisco. He is also the cofounder and chief science advisor at Akili Interactive Labs and Sensync, and co-founder and chief scientist at JAZZ Venture Partners. Dr. Gazzaley is a frequent collaborator of former Grateful Dead band member Mickey Hart's, exploring together the intersection of science, tech, art, performance, and gaming.

## KELLY JAKUBOWSKI

Dr. Kelly Jakubowski is a Leverhulme Early Career Fellow in the Department of Music at Durham University. She conducts research on the relationship between music, memories, and emotions.

## PETR JANATA

Dr. Petr Janata is a professor of psychology at the University of California, Davis, and a cognitive neuroscientist. He conducts research on auditory perception and the mechanisms in the brain that allow people to have deep experiences with music.

## HELEN LAVRETSKY

Dr. Helen Lavretsky is a professor-in-residence in the Department of Psychiatry at UCLA and a geriatric integrative psychiatrist specializing in geriatric depression, integrative mental health, and brain suggestibility.

## DANIEL LEVITIN

Dr. Daniel Levitin is a neuroscientist, musician, and author of four *New York Times* bestselling books, including *This Is Your Brain on Music*. His research encompasses music, the brain, health, productivity, and creativity.

## CHARLES LIMB

Dr. Charles Limb is a surgeon, neuroscientist, chief of Otology/Neurotology and Skull Base Surgery, and professor at the University of California, San Francisco. He is an accomplished musician and also serves as the codirector of the Sound Health Network, an initiative that explores the relationship between music and wellness.

## JOANNE LOEWY

Dr. Joanne Loewy is director of the Department of Music Therapy at the Mount Sinai Health System. Her research areas include sedation, assessment, pain, asthma, and NICU music therapy.

## TOD MACHOVER

Tod Machover is a composer, professor, and inventor at the MIT Media Lab, and director of the Media Lab's Opera of the Future group, with a focus on how technology supports and creates innovative music and music performance.

## SUSAN MAGSAMEN

Susan Magsamen is the executive director and founder of the International Arts + Mind Lab (IAM Lab), part of the Brain Science Institute (BSi) at Johns Hopkins University. She is the codirector of NeuroArts Blueprint, which explores the benefits and appeal of music on the neural level.

## RAFAEL PELAYO

Dr. Rafael Pelayo is a sleep specialist at the Stanford Sleep Medicine Center and a clinical professor at the Division of Sleep Medicine in the Department of Psychiatry and Behavioral Science at Stanford University.

## HEATHER READ

Dr. Heather Read is a professor of psychology at the University of Connecticut whose research focuses on sensory neuroscience and sound perception. She is also director of the Brain Computer Interface Group and a research affiliate of the MIT Media Lab.

## DIANA SAVILLE

Diana Saville is the cofounder and COO of BrainMind, a brain-industry accelerator. She is also cofounder and president of the nonprofit Entrepreneur of Your Own Life.

## LAUREL TRAINOR

Dr. Laurel Trainor is a cognitive psychologist and professor at McMaster University. Her research is focused on the connection between music and science in cognition, perception, and children and infant bonding.

# ARTISTS

## INTERVIEWED FOR THIS BOOK

### LAURIE ANDERSON

Laurie Anderson is an avant-garde artist, composer, musician, and film director whose work spans performance art, pop music, and multimedia projects. Known for her integration of technology and art, Anderson was the first artist-in-residence at NASA.

### STEVE AOKI

Steve Aoki is a Grammy-nominated American deejay, record producer, music programmer, and record executive. Aoki is a brain enthusiast and founder of the Aoki Foundation to support brain science and research.

### DAVID BYRNE

David Byrne is a singer, songwriter, musician, film director, producer, author, lecturer, photographer, and visual artist. He is a founding member of the legendary band Talking Heads. He has been honored with an Academy Award, a Golden Globe, and Grammy Awards, among others, and was inducted into the Rock & Roll Hall of Fame with Talking Heads in 2002.

### CITIZEN COPE
(AKA CLARENCE GREENWOOD)

Citizen Cope is an American songwriter, producer, and performer. His music is commonly described as a mix of blues, soul, folk, and rock, and his compositions have been recorded by Carlos Santana, Dido, Pharoahe Monch, and Richie Havens.

### SHEILA E.
(AKA SHEILA ESCOVEDO)

Sheila E. is a trailblazing percussionist, singer, and drummer who became a household name for a lifelong solo career and her many collaborations with Marvin Gaye, Herbie Hancock, and Prince, among countless others. The "percussive powerhouse" was presented with the Latin Grammy Lifetime Achievement Award with her father in 2021.

### MICK FLEETWOOD

Mick Fleetwood is best known as the drummer, cofounder, and leader of legendary two-time Grammy Award-winning rock band Fleetwood Mac. Fleetwood was inducted into the Rock & Roll Hall of Fame with Fleetwood Mac in 1998.

### PETER GABRIEL

Peter Gabriel is an English singer-songwriter, musician, humanitarian, activist, and two-time inductee into the Rock & Roll Hall of Fame. He was originally a cofounder and lead singer of the rock band Genesis, before embarking on an iconic career as a solo artist. He is a prominent advocate for various social and political causes, and a well-known voice in the neuroscience community.

### KOOL
(AKA ROBERT BELL)

Kool is one of the founding members of the legendary Kool & the Gang, winners of seven Grammy Awards and numerous American Music Awards. One of the most sampled band of all time, Kool & the Gang has sold more than 80 million albums worldwide and influenced generations of musicians with 25 Top Ten R&B hits, 9 Top Ten Pop hits, and 31 gold and platinum albums.

### LACHANZE
(AKA RHONDA SAPP)

LaChanze is an actress, singer, and dancer. She won the Tony Award for Best Actress in a Leading Role in a Musical in 2006 for her portrayal of Celie Harris Johnson in *The Color Purple*.

### BRANFORD MARSALIS

Branford Marsalis is a saxophonist, composer, and bandleader. The three-time Grammy Award winner performs with his band as the Branford Marsalis Quartet, and also plays frequently as a soloist with classical ensembles. He was the bandleader of the Tonight Show Band from 1992 to 1995.

### NICK MASON

Nick Mason is a drummer and founding member of the iconic, Grammy Award-winning rock band Pink Floyd. He is the only member to feature on every Pink Floyd album, and the only constant member since the band's formation in 1965. Mason was inducted into the Rock & Roll Hall of Fame with Pink Floyd in 1996.

### ADAM MASTERSON

Adam Masterson is a singer/songwriter from the UK. His critically acclaimed debut album, *One Tale Too Many* let to sharing the stage with legendary performers such as Tori Amos, Amy Winehouse, The Stereophonics, Mick Jones of The Clash, and Patti Smith.

### MOVER
(AKA JONATHAN MOVER)

Mover is a world-renowned professional drummer and percussionist. He is a former member of the bands Marillion and GTR, and has recorded and performed with artists including Aretha Franklin, Alice Cooper, The Tubes, Shakira, Fuel, Everlast, Frank Gambale, Joe Satriani, and Mick Jagger. Mover recently founded Progject, a prog rock all-star band, with some of the genre's most influential musicians.

### YOUSSOU N'DOUR

Youssou N'Dour is a pioneering, Grammy Award-winning Senegalese singer, songwriter, and bandleader. He is a leading proponent of world music, combining traditional music from his homeland with Western popular culture, Cuban rhythms, and contemporary instrumentation.

### QUESTLOVE
(AKA AHMIR THOMPSON)

Questlove is a six-time Grammy Award-winning drummer, deejay, producer, Oscar-winning filmmaker, *New York Times* bestselling author, and cofounder of hip-hop group The Roots. He is also the musical director of *The Tonight Show Starring Jimmy Fallon*.

### HANS ZIMMER

Hans Zimmer is a film score composer whose work has been featured in more than 150 projects. Zimmer has 12 Oscar nominations, for films including *Inception, Rain Man, Gladiator*, and *As Good as It Gets*, while winning Best Original Score for Disney's animated *The Lion King* and for *Dune*.

# GLOSSARY

**acetylcholine** A neuromuscular neurotransmitter involved in REM sleep, memory, and learning

**adenosine** A neurotransmitter that accumulates during waking hours, ultimately resulting in the feeling of sleepiness

**Alzheimer's disease** A disorder that progressively lays waste to memory and cognitive ability

**amygdalae** Two structures, one in each hemisphere, that provide the emotional component of decision-making (singular: amygdala)

**anterior insula** A larger part of the insular cortex responsible for self-awareness

**anxiety** A normal stress reaction that boosts our alertness in situations of uncertainty

**associative memory** Memories associated with connections and relationships between thoughts

**auditory nerve** A group of nerve fibers that send sound from each cochlea to the brain

**autism spectrum disorder (ASD)** A complex neurodevelopmental disorder characterized by deficits in social interaction and behavior, and high sensitivity to sensory stimuli

**autonomic nervous system (ANS)** Responsible for bodily functions and unconscious physical responses to stimuli. Divided into two main branches:

- **parasympathetic nervous system**: regulates activities occurring while the body is at rest

- **sympathetic nervous system**: excites the fight-or-flight response

**axon** A single long trunk of a neuron that communicates information out to other neurons' dendrites

**beats per minute (BPM)** The number of beats measured in one minute; see: tempo

**binaural beat/sound** Two slightly different tones presented to right and left ears at the same time, creating a new tone that becomes a binaural beat

**brain plasticity** The ability of the brain to change and "rewire" in response to external influence

**brain stem** The base of the brain that connects to the spinal cord. It includes the medulla oblongata, pons, and midbrain.

**brain wave entrainment** The specific form of spontaneous synchronization that consists of brain waves matching an external stimulus

**brain waves** Electrical impulses generated by activity in the brain. There are five states of brain waves: gamma waves, beta waves, alpha waves, theta waves, delta waves. See page six for more detail.

**Broca's area** A region of the left frontal cortex first identified by neuroanatomist Paul Broca in 1861. Responsible for speech production and grammatical structure

**cerebellum** A primitive region located behind the brain stem that coordinates muscle movement, posture, and balance

**cerebral cortex** The outermost layer of the brain, folded into its characteristic wrinkled structure

**cingulate cortex** A part of the cerebral cortex used in judging conflict and error

**cingulate gyrus** Part of the cingulate cortex involved in pain perception, emotional stimuli, and memories

**circadian rhythm** The body's 24-hour biological clock that coordinates with daylight to influence when one feels awake or tired

**consolidation** The second of four stages in memory storage; a natural process that gets rid of unimportant things that the senses pick up throughout the day

**corpus callosum** A large cluster of nerve fibers that connects the brain's left and right hemispheres

**cortisol** A hormone that converts fatty acids into energy available to muscles

**decibel** A measurement of the volume of sound

**default mode network** A system in the brain that is associated with actions of introspection and reflection on identity

**dementia** A disorder that deteriorates mental processes and impacts daily functioning

**dendrite** A neuronal branch that reaches out to receive information from an axon

**depression** A mood disorder that can cause feelings of extreme sadness, dejection, and/or grief.

**dopamine** The brain's key neurotransmitter responsible for the good feelings we get after doing something pleasurable

**dorsolateral prefrontal cortex** A subsection of the frontal lobes that handles inhibitions

**earworm** A song that gets stuck in your head on repeat

**electroencephalography (EEG)** An imaging technology that measures the brain's electrical activity through electrodes on the scalp

**emotions** The body's automatic responses to positive and negative environmental stimuli

**emotional intelligence (EI)** The skill of managing and identifying our and others' feelings

**emotional quotient (EQ)** A proposed expression of the degree of one's emotional intelligence (EI), akin to how IQ measures intelligence

**encoding** The first of four stages in memory storage. Here a memory is initially captured and temporarily laid down as a sensory pattern in the hippocampus.

**endorphins** Hormones that inhibit responses to pain and/or discomfort

**episodic memory** Long-term memories of past personal experiences

**feelings** The brain's interpretations of emotional responses

**fight-or-flight** Instinctive threat-detection response that activates the sympathetic nervous system to precipitate quick and decisive action

**flow state** Also known as being "in the zone," a mental state where one is completely immersed in an activity

**frequency** The rate at which a wave repeats a full cycle

**frisson** Chills or goose bumps induced by a musical experience

**frontal lobe** The foremost region of the brain controlling speech, personality, judgment, problem-solving, and voluntary movement

**functional magnetic resonance imaging (fMRI)** An imaging technology that monitors brain activity due to changes in blood oxygenation. Useful for mapping which brain regions are activated in various conditions

**gray matter** The pinkish-gray outer layer of brain tissue that contains the majority of neurons. It's responsible for bringing sensory information from cells and sensory organs to the brain regions that process sensory information and muscle control.

**harmony** The musical result of two or more notes sounding at the same time, forming an interval (two notes) or a chord (three or more notes)

**hippocampi** Two structures located beneath the cerebral cortex, one in each hemisphere, crucial to forming new memories and governing spatial awareness (singular: hippocampus)

**hormones** Chemicals produced by the endocrine glands that travel throughout the body and regulate various functions such as growth, metabolism, and behavior

**hypnagogia** The state in between wakefulness and sleep that can induce hallucinations, lucid dreaming, and sleep paralysis

**iso principle** A technique used in music therapy where a therapist first matches music with their client's mood, then guides the transition of the mood with appropriate transitions of the music

**key** The group of notes that comprise the harmonic foundation of a musical composition

**limbic system** A collection of brain structures that houses the amygdalae, hippocampi, hypothalamus, cingulate gyrus, basal ganglia, and thalamus. Drives emotional responses, motivation, and instinctive behaviors

**lucid dreaming** The ability to control and guide oneself to new experiences within a dream

**major system** A memory technique involving pairing digits with a set of consonant sounds

**medulla oblongata** A structure in the brain stem that maintains subconscious activities like breathing, heart rate, and sleep, as well as involuntary actions like sneezing and coughing

**melatonin** A hormone that regulates sleep and wakefulness

**melody** A combination of pitch and rhythm that creates a series of musical tones

**motor cortex** Region of the cerebral cortex that controls voluntary muscle movement

**music therapy** The use of music to achieve individualized therapeutic goals to improve mental health

**neocortex** A part of the cerebral cortex used in sensory perception and cognition

**neural firing** The electrical impulses through which neurons communicate with one another

**neural network** A grouping of neurons connected by synapses

**neurodiversity** A movement representing a shift in thought on mental illness. It suggests that a variety of psychiatric disorders, such as autism, are not pathological, but rather natural variations in the genome that deserve wider social acceptance.

**neurons** The primary cells of the nervous system, composed typically in a tree-like structure with one long axon and many branch-like dendrites, which communicate electrochemical signals propagating from the axon of one neuron to the dendrites of others across the synapses, or gaps, between them

**neuroplasticity** The brain's ability to constantly change as new stimuli are presented to it

**neurotransmitters** Brain chemicals of various types that facilitate communication among neurons by carrying electrochemical messages across the synaptic cleft

**noradrenaline** A stress hormone released by the sympathetic nervous system in times of trauma

**norepinephrine** A neurotransmitter that helps prepare the brain and body for fight-or-flight action

**nucleus accumbens** A reward structure in the basal forebrain that regulates the flow of serotonin and dopamine

**oxytocin** A stress-relieving neurotransmitter and hormone that creates feelings of warmth, bonding, security, and trust

**parietal lobes** Two regions of the central cerebral cortex that handle taste, touch, and body awareness

**Parkinson's disease** A neurodegenerative disorder that progressively impairs movement as dopamine-producing neurons die

**perfect pitch** The ability of a person to identify or intone a musical pitch without the reference of another

**pitch** A characteristic of music that describes how high or low a note is

**post-traumatic stress disorder (PTSD)** A debilitating condition resulting from a physical change in the brain because of severe and/or prolonged abuse or exposure to traumatic environmental stimuli

**prefrontal cortex** The front-most part of the frontal cortex that handles inhibitions, higher-order cognition, behaviors, and emotions

**procedural memory** A type of long-term memory that stores the ability to perform tasks and skills. Also known as "muscle memory"

**psychoacoustics** Scientific studies of psychological responses to sound

**reverberation** The persistence of a sound after its source has stopped, caused by repeated reflection of the sound within a closed space. Also, an awesome book

**rhythm** The patterns of silence, sound, and emphasis in a musical composition

**scale** An arrangement of musical notes in order of ascending or descending pitch

**serotonin** A neurotransmitter that regulates mood and social behavior, making you feel calm and relaxed

**shared affective motion experience (SAME)** A model that proposes that our brains process music as intentional human motor actions, not simply random patterns

**working memory** Also known as short-term memory, a severely limited process (15 seconds to a minute) that allows temporary storage of small packets of information

**social baseline theory** A phenomenon in which an interconnected community starts basing decisions on a prediction of continued mutual support

**social flow** When multiple people experience flow states at the same time

**spontaneous synchronization** An effect where oscillations from one system match the oscillation pattern of another

**stress** Strain, tension, or pressure that causes the release of the hormone cortisol in the brain and body. Stress has been found to hinder the growth of new brain cells, shrink the hippocampus, and impede learning.

**stroke** A "brain attack" characterized by a slowing of blood flow to the brain or an internal hemorrhaging from a ruptured aneurysm

**synapses** Neuronal junctions between which neurotransmitters and electrical impulses are transmitted

**synaptic pathway** The connection created between neurons, through synapses

**synesthesia** A sensory phenomenon where inputs from one sense are mistakenly interpreted by a different sense

**tempo** The speed of a musical composition, measured in beats per minute

**temporal lobe** Region of the cerebral cortex responsible for hearing, guilt, emotion, and memory

**timbre** A characteristic of music that describes its tone and sound quality

**tinnitus** A condition in which one most commonly hears a high-pitched ringing in the ears

**Wernicke's area** A region in the rear part of the temporal lobe first identified by neuroanatomist Carl Wernicke in 1874. Responsible for naming things and understanding what others are saying

**white matter** Groups of myelin-wrapped axons and nerve fibers that connect regions of the cerebral cortex, aka gray matter, to one another

# BIBLIOGRAPHY

## CHAPTER 1

"How (and Why) to Boost Your Alpha Brainwaves." CABA - the Charity Supporting Chartered Accountants' Wellbeing. March 3, 2017. https://www.caba.org.uk/help-and-guides/information/how-and-why-boost-your-alpha-brainwaves.

Hammond, Claudia. "Why It's Good to Let Your Mind Wander." BBC Reel, 4:03, July 26, 2021. https://www.bbc.com/reel/video/p09mtl6p/why-it-s-good-to-let-your-mind-wander.

University of Nevada, Reno. "Releasing Stress through the Power of Music." University of Nevada, Reno. 2006. https://www.unr.edu/counseling/virtual-relaxation-room/releasing-stress-through-the-power-of-music.

"Popular Songs." GetSongBPM. https://getsongbpm.com/songs.

Cafasso, Jacquelyn. "Do Binaural Beats Have Health Benefits?" Healthline. Healthline Media. October 6, 2017. https://www.healthline.com/health/binaural-beats#potential-benefits.

Weiland, Tracey J, George A Jelinek, Keely E Macarow, Philip Samartzis, David M Brown, Elizabeth M Grierson, and Craig Winter. "Original Sound Compositions Reduce Anxiety in Emergency Department Patients: A Randomised Controlled Trial." Medical Journal of Australia 195 (11-12): 694–98. https://doi.org/10.5694/mja10.10662.

Knight, W. E. J., and N. S. Rickard. "Relaxing Music Prevents Stress-Induced Increases in Subjective Anxiety, Systolic Blood Pressure, and Heart Rate in Healthy Males and Females." Journal of Music Therapy 38 (4): 254–72. https://doi.org/10.1093/jmt/38.4.254.

Goldsby, Tamara L., Michael E. Goldsby, Mary McWalters, and Paul J. Mills. "Effects of Singing Bowl Sound Meditation on Mood, Tension, and Well-Being: An Observational Study." Journal of Evidence-Based Complementary & Alternative Medicine 22 (3): 401–6. https://doi.org/10.1177/2156587216668109.

MacMillan, Amanda. "Why Nature Sounds Help You Relax, according to Science." Health.com. April 5, 2017. https://www.health.com/condition/stress/why-nature-sounds-are-relaxing.

Hadhazy, Adam. "Why Does the Sound of Water Help You Sleep?" Livescience.com. Live Science. January 18, 2016. https://www.livescience.com/53403-why-sound-of-water-helps-you-sleep.html.

Rogers, Ann E. "The Effects of Fatigue and Sleepiness on Nurse Performance and Patient Safety." Nih.gov. Agency for Healthcare Research and Quality (US). April 2008. https://www.ncbi.nlm.nih.gov/books/NBK2645/.

Francis, Enjoli. "Sleep Deprivation Blamed for JetBlue Pilot's March Meltdown." ABC News, July 10, 2012. https://abcnews.go.com/Blotter/Tired_Skies/sleep-deprivation-blamed-jetblue-pilot-clayton-osbons-march/story?id=16751079.

"Can Daylight Saving Time Hurt the Heart? Prepare Now for Spring." Heart.org. 2018. https://www.heart.org/en/news/2018/10/26/can-daylight-saving-time-hurt-the-heart-prepare-now-for-spring.

Reddy, Sujana, Sandeep Sharma. 2018. "Physiology, Circadian Rhythm." Nih.gov. StatPearls Publishing. October 27, 2018. https://www.ncbi.nlm.nih.gov/books/NBK519507.

Suni, Eric. "Sleep for Teenagers." Sleep Foundation. Sleep Foundation. August 5, 2020. https://www.sleepfoundation.org/teens-and-sleep.

Better Health Channel. "Teenagers and Sleep." Better Health Channel. Victoria State Government. 2018. https://www.betterhealth.vic.gov.au/health/healthyliving/teenagers-and-sleep.

"Sleep/Wake Cycles." Hopkins Medicine. https://www.hopkinsmedicine.org/health/conditions-and-diseases/sleepwake-cycles.

Milliken, Grennan. "Your brain stays half-awake when you sleep in a new place." Popular Science. April 21, 2016. https://www.popsci.com/your-brain-stays-half-awake-when-you-sleep-in-new-place.

Harmat, László, Johanna Takács, Róbert Bódizs. "Music Improves Sleep Quality in Students." Journal of Advanced Nursing. May 1, 2008. https://pubmed.ncbi.nlm.nih.gov/18426457.

RMIT University. "Sound of music: How melodic alarms could reduce morning grogginess: The sounds that wake you up could be affecting how groggy and clumsy you are in the morning." ScienceDaily. February 3, 2020. www.sciencedaily.com/releases/2020/02/200203104505.htm.

Donovan, Jim. "How to Trick Your Brain into Falling Asleep | Jim Donovan | TEDxYoungstown." TEDx, 12:27, September 2018. https://www.ted.com/talks/jim_donovan_how_to_trick_your_brain_into_falling_asleep.

Suni, Eric. "Stages of Sleep." Sleep Foundation. August 14, 2020. https://www.sleepfoundation.org/how-sleep-works/stages-of-sleep.

"Sleep on It." 2017. NIH News in Health. July 13, 2017 https://newsinhealth.nih.gov/2013/04/sleep-it.

"What Happens in the Brain during Sleep?" 2015. Scientific American Mind 26 (5): 70–70. https://doi.org/10.1038/scientificamericanmind0915-70a.

Coombes, Elizabeth. "Anxiety: A Playlist to Calm the Mind from a Music Therapist." The Conversation. November 22, 2019. https://theconversation.com/anxiety-a-playlist-to-calm-the-mind-from-a-music-therapist-121655.

Aggarwal-Schifellite, Manisha. "Research Shows Lullabies in Any Language Relax Babies." 2020. Harvard Gazette. October 19, 2020. https://news.harvard.edu/gazette/story/2020/10/research-shows-lullabies-in-any-language-relax-babies/.

Castro, Joseph. "What Is White Noise?" Live Science. July 29, 2013. https://www.livescience.com/38387-what-is-white-noise.html.

Newsom, Rob. "Music and sleep." Sleep Foundation. OneCare. June 24, 2021. https://www.sleepfoundation.org/noise-and-sleep.

Passman, Jordan. "The World's Most Relaxing Song." Forbes. November 23, 2016. https://www.forbes.com/sites/jordanpassman/2016/11/23/the-worlds-most-relaxing-song/?sh=538f2fa02053.

## CHAPTER 2

Hawks, John. "How Has the Human Brain Evolved over the Years?" Scientific American Mind 24 (3): 76–76. https://doi.org/10.1038/scientificamericanmind0713-76b.

Kwong, Emily. "Understanding Unconscious Bias" Interview with Pragya Agarwal. Short Wave. Podcast audio, 12:40. July 15, 2020. https://www.npr.org/2020/07/14/891140598/understanding-unconscious-bias.

Bailey, Chris. "How to Get Your Brain to Focus | Chris Bailey | TEDxManchester." TEDx. YouTube video, 15:56. April 5, 2019. https://www.youtube.com/watch?v=Hu4Yvq-g7_Y&list=RDLVHu4Yvq-g7_Y&start_radio=1&t=12s.

Trafton, Anne. "How We Tune Out Distractions." MIT News. Massachusetts Institute of Technology. June 12, 2019. https://news.mit.edu/2019/how-brain-ignores-distractions-0612.

Mark, Gloria, Shamsi Iqbal, Mary Czerwinski, Paul Johns, and Akane Sano. "Neurotics Can't Focus: An in Situ Study of Online Multitasking in the Workplace." Microsoft.com. May 1, 2016. https://www.microsoft.com/en-us/research/publication/neurotics-cant-focus-an-situ-of-online-multitasking-in-the-workplace/.

Lanese, Nicoletta. "Is There Actually Science Bhind 'Dopamine Fasting'?" Livescience.com. November 19, 2019. https://www.livescience.com/is-there-science-behind-dopamine-fasting-trend.html.

Witte, Martina de, Anouk Spruit, Susan van Hooren, Xavier Moonen, and Geert-Jan Stams. "Effects of Music Interventions on Stress-Related Outcomes: A Systematic Review and Two Meta-Analyses." Health Psychology Review 14 (2): 1–31. https://doi.org/10.1080/17437199.2019.1627897.

Mehta, Ravi, Rui (Juliet) Zhu, and Amar Cheema. "Is Noise Always Bad? Exploring the Effects of Ambient Noise on Creative Cognition." Journal of Consumer Research 39 (4): 784–99. https://doi.org/10.1086/665048.

Hurless, Nicole, Aldijana Mekic, Sebastian Peña, Ethan Humphries, Hunter Gentry, and David Nichols. "Music Genre Preference and Tempo Alter Alpha and Beta Waves in Human Non-Musicians." 2013. https://impulse.appstate .edu/sites/impulse.appstate.edu/files/Hurless%20et%20al.%20_0.pdf.

Kučikiené, Domanté, Rūta Praninskiené. "The impact of music on the bioelectrical oscillations of the brain." Acta medica Lituanica vol. 25,2 (2018): 101–106. https://www.ncbi.nlm.nih.gov/pmc/articles/ PMC6130927.

American Associates, Ben-Gurion University of the Negev. "Favorite music makes teens drive badly: Teen driver music preferences increase errors and distractibility." ScienceDaily. www.sciencedaily.com/releases/ 2013/08/130823091347.htm.

Schenkman, Lauren. "In the Brain, Seven Is a Magic Number." ABC News, November 27, 2009. https://abcnews.go.com/Technology/ brain-memory-magic-number/story?id=9189664.

Gorlick, Adam. "Media Multitaskers Pay Mental Price, Stanford Study Shows." Stanford University. August 24, 2009. https://news.stanford.edu/ news/2009/august24/multitask-research-study-082409.html.

Hammond, Claudia. "Does Listening to Mozart Really Boost Your Brainpower?" BBC Future. 2014. https://www.bbc.com/future/ article/20130107-can-mozart-boost-brainpower.

Raypole, Crystal. "Here's How Music Can Help You Concentrate." Healthline. July 29, 2020. https://www.healthline.com/health/ does-music-help-you-study#benefits-of-music-for-studying.

Kusnierek, Timmy. "New Study Confirms Listening to EDM at Work Can Make You a Better Employee." 2016. Your EDM. June 4, 2016. https://www. youredm.com/2016/06/04/study-proves-edm-boosts-your- productivity/#:~:text=According%20to%20a%20recent%20study.

Payne, Chris. "NFL Players Talk Music: What's on Their Pre-Game Playlists?" Billboard. September 4, 2014. https://www.billboard.com/ articles/news/6243644/nfl-player-music-playlist-football-durant- schwartz-fluker-jenkins-cole-bouye.

"MP3 to BPM (Song Analyser)." GetSongBPM.com. 2019. https://getsongbpm.com/tools/audio.

"Potential Project - Focused Minds, Organizational Excellence." Potential Project. https://www.potentialproject.com/.

Chodosh, Sara. "You should be listening to video game soundtracks at work." Popular Science. January 26, 2018. https://www.popsci.com/ work-productivity-listening-music/.

Patel, Deep. "These 6 Types of Music Are Known to Dramatically Improve Productivity." Entrepreneur. January 9, 2019. https://www.entrepreneur .com/article/325492.

Downing, Sam. "Song stuck in your head? Scientists might have figured out how to banish it forever." Nine. 2015. https://www.nine.com.au/ entertainment/viral/rip-earworms.

**CHAPTER 3**

Chilton, Martin. "Deconstructing the Love Song: How and Why Love Songs Work." UDiscover Music. February 14, 2020. https://www.udiscovermusic. com/in-depth-features/deconstructing-the-love-song-how-they-work.

Kest, Noah. "The Science of Music and Love." 34th St. February 13, 2018. https://www.34st.com/article/2018/02/science-of-music-and-love.

Salleh, Anna. "What Science Can Tell Us about the Music of Love," ABC News (Australian Broadcast Corporation). June 6, 2017. https://www.abc.net.au/news/science/2017-06-07/ what-science-can-tell-us-about-the-best-music-for-romancing/8583892.

Pereira, Carlos Silva, João Teixeira, Patrícia Figueiredo, et al. "Music and Emotions in the Brain: Familiarity Matters." Edited by Jay Pillai. PLoS ONE 6 (11): e27241. https://doi.org/10.1371/journal.pone.0027241.

Ifeanyi, K. C. "Your Taste in Music Could Be Ruining Your Relationship." Fast Company. July 31, 2019. https://www.fastcompany.com/90384344/ your-taste-in-music-could-be-ruining-your-relationship.

Boer, Diana, Ronald Fischer, Micha Strack, Michael H. Bond, Eva Lo, and Jason Lam. "How Shared Preferences in Music Create Bonds Between People: Values as the Missing Link." Personality and Social Psychology Bulletin 37, no. 9 (September 2011): 1159–71. https://doi .org/10.1177/0146167211407521.

Titlow, John Paul. "How Music Changes Your Behavior at Home." Fast Company. February 10, 2016. https://www.fastcompany.com/3056554/ how-music-changes-our-behavior-at-home.

Wahl, David. "The Psychology of Listening to Music during Sex." Psychology Today. March 12, 2021. https://www.psychologytoday.com/us/blog/ sexual-self/202103/the-psychology-listening-to-music-during-sex.

Orenstein, Beth. "Music for Low T: Get in Tune with Your Sex Drive." Everyday Health. July 12, 2013. https://www.everydayhealth.com/ low-testosterone/music-for-low-t-get-in-tune-with-your-sex-drive.aspx.

Kuzma, Cindy. "Create the Ultimate Sex Playlist." Men's Health. September 16, 2011. https://www.menshealth.com/sex-women/a19526228/ create-the-ultimate-sex-playlist.

Schäfer, Katharina, Suvi Saarikallio, and Tuomas Eerola. "Music May Reduce Loneliness and Act as Social Surrogate for a Friend: Evidence from an Experimental Listening Study." Music & Science 3 (January): 205920432093570. https://doi.org/10.1177/2059204320935709.

Vann, Madeline. "Can Sad Music Heal Your Broken Heart?" Everyday Health. November 12, 2014. https://www.everydayhealth.com/depression/ can-sad-music-heal-your-broken-heart.aspx.

"Do Ya Think I'm Sexy?" Sonos, Apple Music. February 9, 2016. https:// musicmakesithome.com/post/138963442147/do-ya-think-i-sexy.

"Crazy in Love." Sonos, Apple Music. February 9, 2016. https://musicmakesithome.com/post/138963984527/ translate-deutschland-france-nederland-crazy.

"6 scientific facts about how music influences intimacy." Music Crowns. November 13, 2021. https://www.musiccrowns.org/ tips/6-scientific-facts-about-how-music-influences-intimacy.

"Music's Influence on Sexual Behaviors." Tick Pick. https://www.tickpick. com/musics-influence-on-sexual-behaviors.

**CHAPTER 4**

"How Does Marijuana Produce Its Effects?" National Institute on Drug Abuse. 2018. https://www.drugabuse.gov/publications/research-reports/ marijuana/how-does-marijuana-produce-its-effects.

Growney, Claire. "Earliest References to Music Therapy" The History of Music and Art Therapy. https://musicandarttherapy.umwblogs.org/ music-therapy/earliest-references-to-music-therapy.

"Music Therapy and Military Populations." American Music Therapy Association. 2011. https://www.musictherapy.org/research/ music_therapy_and_military_populations.

Growney, Claire. "Music Therapy during the World Wars." The History of Music and Art Therapy. https://musicandarttherapy.umwblogs.org/ music-therapy/music-therapy-in-the-1900s.

"About the Certification Board for Music Therapists." CBMT. 2019. https:// www.cbmt.org/about.

Novotney, Amy. "Music as Medicine." Apa.org. November 2013. https:// www.apa.org/monitor/2013/11/music.

Loewy, Joanne. "NICU Music Therapy: Song of Kin as Critical Lullaby in Research and Practice." Annals of the New York Academy of Sciences 1337 (1): 178–85. https://doi.org/10.1111/nyas.12648.

Waterhouse, J., P. Hudson, and B. Edwards. "Effects of Music Tempo upon Submaximal Cycling Performance." Scandinavian Journal of Medicine & Science in Sports 20 (4): 662–69. https://doi.org/10.1111/j.1600-0838.2009.00948.x.

Thakare, Avinash E, Ranjeeta Mehrotra, and Ayushi Singh. "Effect of Music Tempo on Exercise Performance and Heart Rate among Young Adults." International Journal of Physiology, Pathophysiology and Pharmacology 9 (2): 35–39. https://www.ncbi.nlm.nih.gov/pmc/articles/PMC5435671.

"10 Percent of US Adults Have Drug Use Disorder at Some Point in Their Lives." National Institutes of Health (NIH). November 18, 2015. https:// www.nih.gov/news-events/news-releases/10-percent-us-adults-have- drug-use-disorder-some-point-their-lives.

Gardstrom, Susan C., and Wiebke S. Diestelkamp. "Women with Addictions Report Reduced Anxiety after Group Music Therapy: A Quasi-Experimental Study." Voices: A World Forum for Music Therapy 13 (2). https://doi.org/ 10.15845/voices.v13i2.681.

Erkkilä, Jaakko, and Tuomas Eerola. "Review of Music Therapy Methods in the Treatment of Gambling Addiction." Music Therapy Today (Online) IV, no.3 (June). https://www.researchgate.net/publication/228092588_ Music_therapy_methods_in_the_treatment_of_gambling_addiction.

"World Federation of Music Therapy." Music Therapy World. http://musictherapyworld.net.

Sol Republic. "Sound Over Pounds: Survey Finds Two Out Of Three People Cut Their Workout Short Or Ditch It Completely Without Headphones." PR Newswire. April 2, 2014. https://www.prnewswire.com/news- releases/sound-over-pounds-survey-finds-two-out-of-three- people-cut-their-workout-short-or-ditch-it-completely-without- headphones-253570611.html.

"Does Synchronizing Brain Waves Bring Harmony?" AlzForum. April 14, 2021. https://www.alzforum.org/news/conference-coverage/ does-synchronizing-brain-waves-bring-harmony.

Jabr, Ferris. "Let's Get Physical: The Psychology of Effective Workout Music." *Scientific American*. March 20, 2013. https://www.scientificamerican.com/article/psychology-workout-music.

**CHAPTER 5**

Suttie, Jill. "Four Ways Music Strengthens Social Bonds." Greater Good. https://greatergood.berkeley.edu/article/item/four_ways_music_strengthens_social_bonds.

Harvey, Alan R. "Links Between the Neurobiology of Oxytocin and Human Musicality." Frontiers in Human Neuroscience. August 26, 2020. https://www.frontiersin.org/articles/10.3389/fnhum.2020.00350/full.

Keeler, Jason R., Edward A. Roth, Brittany L. Neuser, John M. Spitsbergen, Daniel J M Waters, and John-Mary Vianney. "The Neurochemistry and Social Flow of Singing: Bonding and Oxytocin." Frontiers in Human Neuroscience. September 23, 2015. https://www.ncbi.nlm.nih.gov/pmc/articles/PMC4585277.

Fishbein, Adam. "Birds Can Tell Us a Lot about Human Language." Scientific American Blog Network. February 02, 2018. https://blogs.scientificamerican.com/observations/birds-can-tell-us-a-lot-about-human-language.

New Jersey Institute of Technology. "Duetting songbirds 'mute' the musical mind of their partner to stay in sync." ScienceDaily. www.sciencedaily.com/releases/2021/05/210531153207.htm.

Morelle, Rebecca. "Choir Singers 'synchronise Their Heartbeats'." BBC News. July 09, 2013. https://www.bbc.com/news/science-environment-23230411.

Horn, Stacy. "Why Joining a Choir Is the Easiest Way to Make Yourself Happier." Slate. July 25, 2013. https://slate.com/human-interest/2013/07/singing-in-a-choir-research-shows-it-increases-happiness.html

Kreutz, Gunter et al. "Effects of choir singing or listening on secretory immunoglobulin A, cortisol, and emotional state." Journal of Behavioral Medicine 27, no. 6 (2004): 623-35. doi: 10.1007/s10865-004-0006-9.

Weinstein, Daniel, Jacques Launay, Eiluned Pearce, et al. "Singing and Social Bonding: Changes in Connectivity and Pain Threshold as a Function of Group Size." Evolution and Human Behavior. October 19, 2015. https://www.sciencedirect.com/science/article/abs/pii/S1090513815001051.

Dunbar, R. I. M. "Neocortex Size, Group Size, and the Evolution of Language." Current Anthropology 34, no. 2 (1993): 184–93. Doi:10.1086/204160. https://pdodds.w3.uvm.edu/files/papers/others/1993/dunbar1993a.pdf.

Harré, Michael. "Social Network Size Linked to Brain Size." Scientific American. August 07, 2012. https://www.scientificamerican.com/article/social-network-size-linked-brain-size.

Baimel, Adam, Rachel L. Severson, Andrew S. Baron, and Susan A. J. Birch. "Enhancing "theory of mind" through Behavioral Synchrony." Frontiers In. June 23, 2015. https://www.frontiersin.org/articles/10.3389/fpsyg.2015.00870/full.

Overy, Katie. "Making Music in a Group: Synchronization and Shared Experience." Annals of the New York Academy of Sciences 1252, no. 1 (2012): 65–68. doi:10.1111/j.1749-6632.2012.06530.x. https://music.uwo.ca/research/research-groups/mlal/pdf/Overy2012NYAS.pdf.

Seaver, Maggie. "Study Says Going to Concerts Leads to a Longer, Happier Life." Real Simple. https://www.realsimple.com/health/preventative-health/live-music-happiness-study.

Kokal, Idil, Annerose Engel, Sebastian Kirschner, and Christian Keysers. "Synchronized Drumming Enhances Activity in the Caudate and Facilitates Prosocial Commitment - If the Rhythm Comes Easily." PLOS ONE. November 16, 2011. https://journals.plos.org/plosone/article?id=10.1371/journal.pone.0027272.

Schulkin, Jay, and Greta B. Raglan. "The Evolution of Music and Human Social Capability." Frontiers In. September 17, 2014. https://www.frontiersin.org/articles/10.3389/fnins.2014.00292/full.

Morin, Amy. "Can You Die from Loneliness?" Psychology Today. 2019. https://www.psychologytoday.com/us/blog/what-mentally-strong-people-dont-do/201901/can-you-die-loneliness.

Manson, Joshua. "How Many People Are in Solitary Confinement Today?" Solitary Watch. January 4, 2019. https://solitarywatch.org/2019/01/04/how-many-people-are-in-solitary-today.

Raghunathan, Raj. "The Need to Love." Psychology Today. January 8, 2014. https://www.psychologytoday.com/us/blog/sapient-nature/201401/the-need-love.

Schulten, Katherine. "Is 14 a 'Magic Age' for Forming Cultural Tastes?" New York Times. May 25, 2011. https://learning.blogs.nytimes.com/2011/05/25/is-14-a-magic-age-for-forming-cultural-tastes.

Sol Republic. "Sound over Pounds: Survey Finds Two Out of Three People Cut Their Workout Short or Ditch It Completely Without Headphones." PR Newswire. April 2, 2014. https://www.prnewswire.com/news-releases/sound-over-pounds-survey-finds-two-out-of-three-people-cut-their-workout-short-or-ditch-it-completely-without-headphones-253570611.html

University of Montreal. "Singing calms baby longer than talking: New study shows that babies become distressed twice as fast when listening to speech compared to song." ScienceDaily. https://www.sciencedaily.com/releases/2015/10/151028054532.htm.

Associated Press. "New Study Shows Unique Ways Dads Are Bonding w/ Kids." FOX 47. Scripps Local Media, June 13, 2019. https://www.fox47news.com/news/local-news/new-study-shows-unique-ways-dads-are-bonding-w-kids.

**CHAPTER 6**

Guillery-Girard, Bérengère, Sylvie Martins, Sebastien Deshayes, et al. "Developmental trajectories of associative memory from childhood to adulthood: a behavioral and neuroimaging study." Frontiers in Human Neuroscience. September 27, 2013. https://www.frontiersin.org/articles/10.3389/fnbeh.2013.00126/full.

Upile, Tahwinder, Fabian Sipaul, Waseem Jerjes, Sandeep Singh, Seyed Ahmad Nouraei, Mohammed El Maaytah, Peter Andrews, John Graham, Colin Hopper, and Anthony Wright. "The Acute Effects of Alcohol on Auditory Thresholds." BMC Ear, Nose and Throat Disorders 7, no. 1 (September 18, 2007). https://doi.org/10.1186/1472-6815-7-4.

Gradus, Jaimie, and Matthew Friedman. "Research Quarterly Advancing Science and Promoting Understanding of Traumatic Stress PTSD and Death from Suicide Point: PTSD Increases Risk for Death from Suicide." US Department of Veterans Affairs, National Center for PTSD. https://www.ptsd.va.gov/publications/rq_docs/V28N4.pdf.

Landis-Shack, Nora, Adrienne J Heinz, and Marcel O Bonn-Miller. "Music Therapy for Posttraumatic Stress in Adults: A Theoretical Review." *Psychomusicology* 27 (4): 334–42. https://www.ncbi.nlm.nih.gov/pmc/articles/PMC5744879.

Stafford, Tom. "Why Does Walking through Doorways Make Us Forget?" BBC. March 8, 2016. https://www.bbc.com/future/article/20160307-why-does-walking-through-doorways-make-us-forget.

Whitbourne, Susan Krauss "Why and How You Daydream." *Psychology Today*. January 8, 2013. https://www.psychologytoday.com/us/blog/fulfillment-any-age/201301/why-and-how-you-daydream.

Zedelius, Claire. "Daydreaming Might Make You More Creative—but It Depends on What You Daydream about." *Behavioral Scientist*. November 2, 2020. https://behavioralscientist.org/daydreaming-might-make-you-more-creative-but-it-depends-on-what-you-daydream-about.

Warner, Jennifer. "Loud Bar Music Makes You Drink More." WebMD. July 18, 2008. https://www.webmd.com/mental-health/addiction/news/20080718/loud-bar-music-makes-you-drink-more.

Engels, Rutger C. M. E., Gert Slettenhaar, Tom ter Bogt, and Ron H. J. Scholte. "Effect of Alcohol References in Music on Alcohol Consumption in Public Drinking Places." *The American Journal on Addictions* 20 (6): 530–34. https://doi.org/10.1111/j.1521-0391.2011.00182.x.

MacArdle, Mark. "Does Country Music Drink More than Other Genres?" Medium. October 14, 2018. https://towardsdatascience.com/does-country-music-drink-more-than-other-genres-a21db901940b.

Mormann, Nicole. "Spotify Reveals Top Pot Playlists in Honor of 4/20." *Hollywood Reporter*. April 20, 2015. https://www.hollywoodreporter.com/news/music-news/spotify-reveals-top-pot-playlists-790195.

Seibert, Erin. "Let's Talk About Iso-Principle: The Introduction." Music Therapy Time. May 19, 2015. https://musictherapytime.com/2015/05/19/lets-talk-about-iso-principle-the-introduction.

**CHAPTER 7**

"Mozart - One of History's Most Tragic Figures." WNO. August 25, 2020. https://wno.org.uk/news/mozart-one-of-historys-most-tragic-figures.

Hambrick, David. "What Makes a Prodigy?" *Scientific American*. September 22, 2015 https://www.scientificamerican.com/article/what-makes-a-prodigy1.

Lewis, Penelope A., Günther Knoblich, and Gina Poe. 2018. "How Memory Replay in Sleep Boosts Creative Problem-Solving." *Trends in Cognitive Sciences* 22 (6): 491–503. https://doi.org/10.1016/j.tics.2018.03.009.

"Thomas Edison's Secret Trick to Maximize His Creativity by Falling Asleep." HuffPost. October 28, 2017. https://www.huffpost.com/entry/thomas-edisons-secret-trick-to-maximize-his-creativity_b_59f4d276e4b06ae9067ab91c.

Larson, Jennifer. "Alpha Brain Waves: What Are They and Why Are They Important?" Healthline. October 9, 2019. https://www.healthline.com/health/alpha-brain-waves#what-are-they.

Yu, Christine. "Why Do We Get Our Best Ideas in the Shower?" Headspace. https://www.headspace.com/articles/shower-epiphanies.

"Keep Your Brain Young with Music." Johns Hopkins Medicine. https://www.hopkinsmedicine.org/health/wellness-and-prevention/keep-your-brain-young-with-music.

Suttie, Jill. "How Music Helps Us Be More Creative." Greater Good. 2017. https://greatergood.berkeley.edu/article/item/how_music_helps_us_be_more_creative.

Mehta, Ravi, Rui (Juliet) Zhu, and Amar Cheema. "Is Noise Always Bad? Exploring the Effects of Ambient Noise on Creative Cognition." Journal of Consumer Research 39 (4): 784–99. https://doi.org/10.1086/665048.

"What Is Neuroplasticity? Definition & FAQs." Emotiv. https://www.emotiv.com/glossary/neuroplasticity.

Lu, Jing, Hua Yang, Xingxing Zhang, Hui He, Cheng Luo, and Dezhong Yao. "The Brain Functional State of Music Creation: An FMRI Study of Composers." Scientific Reports 5 (1). https://doi.org/10.1038/srep12277.

Lesiuk, Teresa. "The Effect of Music Listening on Work Performance." Psychology of Music 33, no. 2 (April 2005): 173–91. https://doi.org/10.1177/0305735605050650.

"This Is Your Brain on Jazz: Researchers Use MRI to Study Spontaneity, Creativity" Johns Hopkins Medicine. February 26, 2008. https://www.hopkinsmedicine.org/news/media/releases/this_is_your_brain_on_jazz_researchers_use_mri_to_study_spontaneity_creativity.

"The Averaged Inter-Brain Coherence between the Audience and a Violinist Predicts the Popularity of Violin Performance." NeuroImage 211 (May): 116655. https://doi.org/10.1016/j.neuroimage.2020.116655.

"Correlation between Math and Music Ability." Brain Balance Centers. https://www.brainbalancecenters.com/blog/correlation-between-math-and-music-ability.

Wolchover, Natalie. "What Type of Music Do Pets Like?" Live Science. March 19, 2012. https://www.livescience.com/33780-animal-music-pets.html.

Herron, Isaac. "10 Mind-Blowing Music Facts." Youth Time. May 28, 2021. https://youth-time.eu/10-mind-blowing-music-facts.

Cornish, Audie, Robert Siegel, Amy Wang. "When It Comes To CDs In 2016, Mozart Outsells Beyonce, Adele And Drake" Interview with Pragya Agarwal. All Things Considered. Podcast audio, 1:00. December 12, 2016. https://www.npr.org/2016/12/12/505311193/when-it-comes-to-cds-in-2016-mozart-outsells-beyonce-adele-and-drake.

"Spotify Listeners Are Getting Nostalgic: Behavioral Science Writer David DiSalvo and Cyndi Lauper Share Why." Spotify Newsroom. Spotify, April 14, 2020. https://newsroom.spotify.com/2020-04-14/spotify-listeners-are-getting-nostalgic-behavioral-science-writer-david-disalvo-and-cyndi-lauper-share-why.

**CHAPTER 8**

Westmaas, Reuben. "What Getting Chills from Music Says about Your Brain." Discovery. August 1, 2019. https://www.discovery.com/science/Getting-Chills-from-Music.

Jennings, Alistair. "Why Does Music Make Us Emotional?" Inside Science. December 29, 2017. https://www.insidescience.org/video/why-does-music-make-us-emotional.

"'Happy' and 'Sad' Music Differs across Cultures - Durham University." Durham University. January 14, 2021. https://www.dur.ac.uk/news/newsitem/?itemno=43528.

Mohana, Malini. "Music & How It Impacts Your Brain, Emotions." Psych Central. May 17, 2016. https://psychcentral.com/lib/music-how-it-impacts-your-brain-emotions.

Lubin, Gus. "8 Amazing Effects That Background Music Has on Sales." Business Insider. July 21, 2011. https://www.businessinsider.com/effects-of-music-on-sales-2011-7.

Allen, David. "'King Kong' by Max Steiner (1933) and James Newton Howard (2005): A Comparison of Scores and Contexts." February 7, 2014. https://davidallencomposer.com/blog/king-kong-max-steiner-james-newton-howard-comparison.

Roberts, Maddy Shaw. "John Williams Receives His 52nd Oscar Nomination for 'Rise of Skywalker' Score." Classic FM. January 13, 2020. https://www.classicfm.com/composers/williams/52nd-oscar-nomination-star-wars-rise-skywalker.

Daniel, Alex. "40 Facts About Music That Really Sing." Best Life. February 27, 2019. https://bestlifeonline.com/music-facts.

Hartmann, Graham. "Science: Queen's 'Don't Stop Me Now' Is the World's Most Uplifting Song." Loudwire. February 13, 2019. https://loudwire.com/queen-dont-stop-me-now-worlds-most-uplifting-song.

Sweetland Edwards, Haley. "The Best and Worst Political Campaign Songs (but Mostly the Worst)." Mental Floss. Minute Media, February 7, 2019. https://www.mentalfloss.com/article/29066/best-and-worst-political-campaign-songs-mostly-worst.

**CHAPTER 9**

Cherry, Kendra. "Music Preferences and Your Personality." VeryWell Mind. January 28, 2020. https://www.verywellmind.com/music-and-personality-2795424.

Swartz, Aimee. "Do You Have A Mellow Music Brain Or An Intense One?" Popular Science. July 22, 2015. https://www.popsci.com/do-you-have-mellow-music-brain-or-intense-one.

Heshmat, Shahram. "How Music Brings People Together." Psychology Today. November 30, 2021. https://www.psychologytoday.com/us/blog/science-choice/202111/how-music-brings-people-together.

Demmrich, Sarah. "Music as a Trigger of Religious Experience: What Role Does Culture Play?" Psychology of Music 48, no. 1 (January 2020): 35–49. https://doi.org/10.1177/0305735618779681.

McFadyen, Darin. "Neuroscience shows listening to music has kind of the same effect as meditation." Quartz. May 12, 2018. https://qz.com/quartzy/1274667/neuroscience-shows-listening-to-music-has-kind-of-the-same-effect-as-meditation.

Fredrickson, Barbara L., Michael A. Cohn, Kimberly A. Coffey, Jolynn Pek, and Sandra M. Finkel. "Open Hearts Build Lives: Positive Emotions, Induced through Loving-Kindness Meditation, Build Consequential Personal Resources." Journal of Personality and Social Psychology 95, no. 5 (November 2008): 1045–62. https://doi.org/10.1037/a0013262.

Meng, Qi, Tingting Zhao, Jian Kang. "Influence of Music on the Behaviors of Crowd in Urban Open Public Spaces." Frontiers In. April 27, 2018. https://www.frontiersin.org/articles/10.3389/fpsyg.2018.00596/full.

Palmer, A. "Violent song lyrics may lead to violent behavior." Monitor on Psychology 34, no. 7 (July/August 2003). https://www.apa.org/monitor/julaug03/violent.

Ritter, Simone, Sam Ferguson. "Happy creativity: Listening to happy music facilitates divergent thinking." PLoS ONE 12, 9 (2017). https://doi.org/10.1371/journal.pone.0182210.

ter Bogt, Tom, Natale Canale, Michela Lenzi, Alessio Vieno, and Regina van den Eijnden. "Sad Music Depresses Sad Adolescents: A Listener's Profile." Psychology of Music 49, no. 2 (March 2021): 257–72. https://doi.org/10.1177/0305735619849622.

Tayag, Yasmine. "Brain Manipulation Could Help Failing Artists Win Back Fans, Study Suggests." Impact. November 21, 2017. https://www.inverse.com/article/38634-music-taste-brain-manipulation.

Wu, Xiao, and Xuejing Lu. "Musical Training in the Development of Empathy and Prosocial Behaviors." Frontiers in Psychology 12: 661769. May 11, 2021. https://doi.org/10.3389/fpsyg.2021.661769.

Barker, Eric. "Should everyone be required to learn a musical instrument?" Barking Up the Wrong Tree. July 6, 2011. https://www.bakadesuyo.com/2011/07/should-everyone-be-required-to-learn-a-musica.

Stillman, Jessica. "Science: Listening to This Type of Music Makes You a Better Person." Inc. December 4, 2017. https://www.inc.com/jessica-stillman/science-listening-to-this-type-of-music-makes-you-a-better-person.html.

Allen, Summer. "Five Ways Music Can Make You a Better Person." Greater Good. Berkeley, November 14, 2017. https://greatergood.berkeley.edu/article/item/five_ways_music_can_make_you_a_better_person.

Stillman, Jessica. "This Quirky Activity Is the Best Way to Make Friends, According to Science." Inc. May 17, 2017. https://www.inc.com/jessica-stillman/this-quirky-activity-is-the-best-way-to-make-friends-according-to-science.html.

Collingwood, Jane. "Preferred Music Style Is Tied to Personality." PsychCentral. Healthline. May 17, 2016. https://psychcentral.com/lib/preferred-music-style-is-tied-to-personality.

Backus, Fred. "Streaming Surpasses Radio as the Top Way to Listen to Music." CBS News. April 9, 2021. https://www.cbsnews.com/news/streaming-tops-radio-as-the-top-way-to-listen-to-music.

"Fighting Fire with Sound: Wave Extinguisher" ESO. June 10, 2018. https://www.eso.org/blog/firefighters-fight-fires-with-sound.

Graham, Sarah. "True Cause of Whip's Crack Uncovered." Scientific American. Springer Nature America, Inc., May 28, 2002. https://www.scientificamerican.com/article/true-cause-of-whips-crack.

# ACKNOWLEDGMENTS

To Michael Sand and Samantha Weiner, our North Stars.
We can't thank you enough for your trust and your TLC.

And to our colleagues, friends, and other loved ones,
gigantic thanks to each one of you, you mean so much to us.

Stacy Woodruff xo

Rosemary Hermann

Don Hermann

Lois Schwartzman

Adam Masterson

Jaz Masterson

Max Masterson

Jill Gabriel

Melanie Gabriel

Luc Gabriel

Isaac Gabriel

Meabh Flynn

Mike Large

Isabel Brandl

Eddie Walsh

Jason Koonce

Chris Jones

Craig Dubitsky

Lisa Silverman

Diane Shaw

Peggy Garry

Gabby Fisher

Lindsay Mergens

Mamie VanLangen

Monica Shah

Elisa Gonzalez

Abrams Sales Team

Zoë Stone

Michael Thomas

Marc Cimino

Adam Ibarra

Shirley Wu

Marni Condro

Mister Lynch

Tod Machover

Illozoo

Alexis Rosenzweig

Marc Goldberg

Britta Bucholz

Aaron Matusow

Taylor Weekly

James Robinson

Michelle Lahana

Kathey Marsella

Susan Magsamen

Amber Treadway

Laura DiMichele

Shelby Scott

Trish Greene

Lucas Reilly

Ann Marie Wilkins

Victoria Gilbert

Salwa Benloubane

Tony Johnson

Kevin Goins

Angelo Ellerbee

Yvette Alberdingk-Thijm

Patrick Tucker

Martin Mugratsch

Elisa Kim

Scott Braun

Leonard Ssemakula

Annie Ohayan

Toki Wright

Kim Jakwerth

Nikki Kearney

Billy Goldberg

Adam Neuhaus

Joe Sabia

Chris Golier

Bill Campbell

Peter Berkowitz

Mover

Charles Limb

Danielle Parillo

Brenda Ross

Mark Burk

Ahovi Kponou

Peter Haugen

Reyna Mastrosimone

Marc Rosen

Nathan Brackett

Andrew Gaeta

Cory Baker

Rich Antoniello

Greg Werner

Jenna Ruggiero

Jesse Kirshbaum

Alex Kirshbaum

Heather Berlin

The Ticknors

Daymond John

Ted Kingsbery

Alex Hoffman

Rob Bozas

# AND CREDITS

Reverberation Cofounders and Executive Editors
**MICHAEL HERMANN, PETER GABRIEL, ANNA GABRIEL**

Book Design **SAMANTHA MERLEY**

Cover Design **MARC BESSANT**
Interior Illustrations **BRIDGET FITZGERALD**
Chapter Head Illustrations **GAËTAN HEUZÉ**

Consulting Science Editor **DIANA SAVILLE**
Assistant Science Editors **KRITI G. ACHYUTUNI,
RANYA R. BELMAACHI, SHREYA V. KRISHNA**

Lead Researcher **KATIE SEMACK**
Creative Consultant **MEGAN KINGERY**

Legal **ELLIOT SCHAEFFER**
Finance **LENNY SANDER, BILL PELLEGRINO**

**SPECIAL THANKS** Joy Allen, Maayan Levavi, Ryan Schinman,
Universal Music Group, Eric Simon, Paula Kaplan, and to all of the
amazing artists and scientists who let us borrow their brains.

Author **KEITH BLANCHARD**

Editor **SAMANTHA WEINER**
Managing Editor **LISA SILVERMAN**
Production Manager **LARRY PEKAREK**

Library of Congress Control Number: 2022933589
ISBN: 978-1-4197-6189-8
eISBN: 978-1-64700-670-9

Printed and bound in the United States
10 9 8 7 6 5 4 3 2 1

This book is intended for entertainment purposes only, and nothing within should be construed as medical, legal, or any other professional advice. All reasonable attempts have been made to ensure the accuracy of the information contained within, including text and graphical elements, but this is a fast-moving field with competing theories and opinions and other challenges to rigorous verification, and accuracy can't be guaranteed. Always seek the advice of your doctor before changing any health or exercise routine.

Abrams Image books are available at special discounts when purchased in quantity for premiums and promotions as well as fundraising or educational use. Special editions can also be created to specification. For details, contact specialsales@abramsbooks.com or the address below.

ABRAMS The Art of Books
195 Broadway, New York, NY 10007
abramsbooks.com

WICKED COW STUDIOS